This book was
donated by

John Carpenter

Information Analysis Center

The Shape of Powder-Particle Outlines

MATERIALS SCIENCE AND TECHNOLOGY SERIES

Series Editor: **Professor G. W. Greenwood,**
University of Sheffield, UK

1. The Shape of Powder-Particle Outlines
   Arthur E. Hawkins

The Shape of Powder-Particle Outlines

Arthur E. Hawkins

RESEARCH STUDIES PRESS LTD.
Taunton, Somerset, England

JOHN WILEY & SONS INC.
New York · Chichester · Toronto · Brisbane · Singapore

RESEARCH STUDIES PRESS LTD.
24 Belvedere Road, Taunton, Somerset, England TA1 1HD

Marketing and Distribution:

Australia and New Zealand:
Jacaranda Wiley Ltd.
GPO Box 859, Brisbane, Queensland 4001, Australia

Canada:
JOHN WILEY & SONS CANADA LIMITED
22 Worcester Road, Rexdale, Ontario, Canada

Europe, Africa, Middle East and Japan:
JOHN WILEY & SONS LIMITED
Baffins Lane, Chichester, West Sussex, England

North and South America:
JOHN WILEY & SONS INC.
605 Third Avenue, New York, NY 10158, USA

South East Asia:
JOHN WILEY & SONS (SEA) PTE LTD.
37 Jalan Pemimpin 05-04
Block B Union Industrial Building, Singapore 2057

Library of Congress Cataloging-in-Publication Data

Hawkins, Arthur E., 1923–
The shape of powder-particle outlines / Arthur E. Hawkins.
p. cm. — (Materials science and technology series ; 1)
Includes bibliographical references and index.
ISBN 0-86380-142-0 (Research Studies Press). — ISBN 0-471-93878-5 (Wiley)
1. Particles. 2. Powders. 3. Particle size determination.
I. Title. II. Series: Materials science and technology series (Taunton, England) ; 1.
TA418.78.H38 1993
620'.41—dc20 92-44143
CIP

British Library Cataloguing in Publication Data

A catalogue record for this book
is available from the British Library.

ISBN 0 86380 142 0 (Research Studies Press Ltd.)
ISBN 0 471 93878 5 (John Wiley & Sons Inc.)

Typeset by Bernie Nicholls Typesetting, Aylesbury, UK
Printed in Great Britain by SRP Ltd., Exeter

Contents

Preface

HAMLET:
Do you see yonder cloud that's almost in shape of a camel?
POLONIUS:
By the mass, and 'tis like a camel, indeed.
HAMLET:
Methinks it is like a weasel.
POLONIUS:
It is backed like a weasel.
HAMLET:
Or like a whale?
POLONIUS:
Very like a whale.

Some twenty years ago, when working in a large food research laboratory, I began to investigate the behaviour of certain powders. My long experience since then in the practical aspects of this field has led me to write this book, having two main objects in view.

Firstly to provide a readable account of the measurement of powder-particle outline shape which introduces not only the principles, but also the realities of the subject. The principles are interesting, stimulating and intellectually challenging; the realities are essentially unpleasant. It is difficult to measure outline shapes and even more difficult to relate the behaviour of a powder to the outline shapes of the particles which make up that powder.

The second reason for writing this book was that there is an interest in the measurement of outline shape in disciplines other than powder science. Unfortunately, until recently, there was little communication between the workers in different subjects, so that many useful contributions from biology, botany, geography, geology, medicine and others have been overlooked by powder scientists.

It is not possible for me to claim that the reference list in this book is absolutely complete, but it is extensive; it has been researched across disciplines and the comments on the papers concerned should guide the reader about the usefulness (or otherwise) of reading the originals.

A.E. Hawkins
March 1993

1 Introduction

In the introduction to a paper suggesting a new approach to the representation of particle shape, Davies (1975) gives a dozen examples where information on shape is needed to describe the behaviour of particles in both industrial processes and products. Shape is recognised as one of the fundamental properties of powder particles: it affects, for example, powder packing and thus bulk density, porosity, permeability and cohesion among other properties. The need for describing and measuring particle shape is taken for granted in this book, which is primarily about describing mathematically the two-dimensional outline, silhouette, profile or projection of small particles generally seen through a microscope. The limitations of this description should be appreciated. An analogy is the manner in which a hilly skyline changes in appearance as it is approached; the relation between the observer and overlapping features changes. The apparent outline of a particle can change markedly depending on the viewpoint. This is obvious if the particle is flaky like mica, but is easily overlooked with chunky particles.

Both presentation and sampling are considered in the book. It is possible to expend much time and effort on characterising particles from a powder; but, if the particles are not representative of the bulk or the outlines not representative of the particles, the endeavour is to little purpose.

Part of the book considers the literature describing and using apparently simple methods of particle-outline description by which it should be possible to characterise large numbers of particles in a relatively short time. These rapid methods might be thought to be concomitant with satisfactory sampling of a bulk powder as compared with elaborate descriptions involving a plethora of data slow to collect and deal with (even using a powerful computer facility) and consequently obtained for relatively few particles. This is not necessarily so, and the literature of the latter methods is described in some depth.

It is important to note that some writers use ‘shape of a particle’ as a shorthand when the discussion is about the two-dimensional outline or silhouette of a particle image. When reading some papers, one has the uncomfortable feeling that the authors think of the two concepts as the same. If the reader visualises the two solids of revolution that would be generated if an outline were rotated about two axes at

Fig. 1.1 The appearance and outline shape of an object rotated about a horizontal axis in front of a camera

right-angles, this misfortune should be avoided. For example, in Fig. 1.2 a rectangle is shown to give a long cylinder if it is rotated about its midpoint x-axis and a cylindrical disc if it is rotated about its midpoint y-axis.

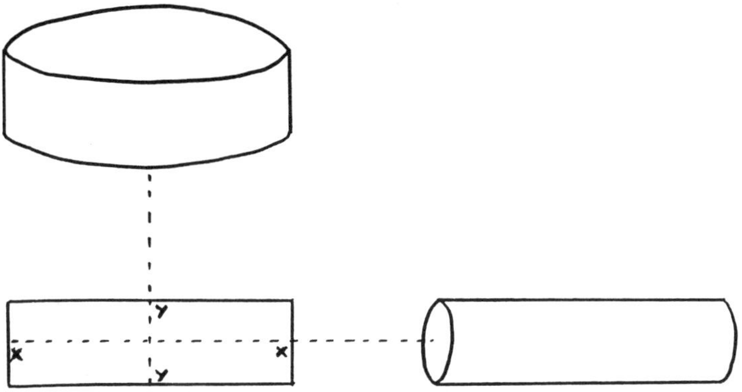

Fig. 1.2 Two solids of revolution that can be generated from a rectangle

Outline shape description is far from being a new topic, even though some of the literature in powder technology gives that impression. Before the First World War, Free (1911) was writing about shape and drag in wind and water relating to soil movement, Sorby (1880) involved shape and surface character in a classification of sand grains into five groups, and Markwick (1937) cites von Rittinger (1867) as detailing shape on the basis partly of principal dimensions. Sixty years ago, Tickell (1931) wrote:

'The examination of mineral fragments, whether in bulk or in grain, has claimed the attention of many workers in the applied physical sciences. Among those exhibiting interest in the subject, and contributing largely to it, are mineralogists, geologists, ceramists, hydrologists, petroleum engineers, metallurgists and chemical engineers.'

In addition to a continuing use by these specialists, one sees today the successful application of two-dimensional shape characterisation in such disparate activities as medicine, biology and botany, to say nothing of military, banking and crime prevention usage.

In only some of these areas has outline shape been related to some behaviour, and, even in powder technology, the need to measure and record powder-particle outlines is not always because some bulk behaviour of the powder is involved. For example, some aspect of identification can be a need. There can be a legal requirement to locate without doubt the source of a pollutant; there can be a change of particle shape at some stage of a manufacturing process, or a requirement to duplicate a particle shape by some alternative process to that in use at present. This is an important consideration because the ultimate use does affect the detail of outline-shape description needed; and the literature in powder technology does emphasise and concentrate on the relation of particle shape to bulk behaviour.

It is worth quoting from a recent book by Svarovsky (1987):

'...it is interesting to note that there is an imbalance in the area of powder testing in favour of very detailed particle size and shape measurements using very sophisticated and expensive laboratory instrumentation, without yet having the fundamental correlations to translate the measured data into secondary behaviour.'

Svarovsky is perhaps touching here on the needs of academic pedagogues compared with those of a more pragmatic nature of the industrial engineer or scientist. The quotation echoes Barrett (1980), who commented that the seductive ease with which measurements of form and roundness can be made is in contrast to the difficulties encountered in ascribing a (geological) meaning to them. Years earlier, Hausner (1966) suggested that it would be worth while investigating the properties of several powders consisting of fairly uniform-sized particles but differing with respect to particle shape: properties such as apparent and tap density, flow and adsorption: and to correlate them with the three shape factors his paper described. The proposal is obviously valid and could be made for any system of shape description and could include other bulk properties, but the effort and funding required to do the work would be considerable. The preparation of 'identical' material of different shapes would be extremely difficult if not impossible, which would mean, in turn, performing a large number of experiments to satisfy the requirements of the statistical treatments to eliminate the unwanted variable(s) from the analysis of results. The magnitude of such a task becomes greater the more detailed the shape description.

It will be seen in a later Chapter that much intellectual effort has been put into finding mathematical descriptions of particle outlines and into the manipulation of equations to condense the detail. The product is, generally, tens of numbers, which, together, describe the particle outline and, in some cases, permit instruments to draw an approximation of the particle outline. There is now a suggestion that such blanket shape descriptors are not necessarily the answer to problems of particle behaviour. After some 15 years of intense effort, it has been concluded by two well-known workers that one cannot measure all the shape features and hope to find the few that are relevant to a given problem. After this long period of intensive publication of detailed mathematical descriptions of outline shapes, epitomised well by Meloy (1980), Meloy & Clark (1986) state that this is not the correct approach and emphasise that there is no universal method for measuring the relevant shape properties and also that there are few useful methods of measuring only relevant shape parameters. They are scathing about most investigators knowing little about the subject and being either unwilling or unable to learn the field. They claim that large amounts of research money are wasted on making shape measurements resulting in no usable shape characterisation data. The authors say in this and other recent publications that one must use intelligence and experience to work out a relevant shape characteristic and then find a method to measure it.

This fresh view about particle shape is put clearly by Drescher et al (1990):

'the current state in shape analysis is distinguished by a number of characterization methods, but the great variety of specific shapes complicates the selection of parameters that are relevant for a particular problem. Therefore, the preferred approach is to characterize single particles 'free of presupposition' and to select technologically relevant parameters using cluster and discriminance algorithms.'

Winkelmolen (1982) said much the same thing as the previously cited authors:

'Sedimentological processes are based on physical and chemical laws. The only way fully to understand the effects of such laws on sediment particles is to describe the grains in parameters which are relevant to the process.'

The new stance is a spur to look at the history of the subject across disciplines so that ideas displaced or lost in the past are considered again in the new climate. However, there is a paucity of references in powder-technology literature; and many do not provide a title, so that sensible selection for reading becomes impossible. A list giving only the author and a journal reference was adequate for the reader whose library carried copies of all the journals cited, but this was rarely the situation in industrial libraries nor, indeed, in the libraries of some universities and research institutes. Today, according to two recent reports, it is not the case in previously well supplied institutions (Erens, 1991; Finch & North, 1991). Many more research workers now need to obtain copies of papers from the British Lending Library, and the waste of time, effort and money on irrelevant references can be substantial. One purpose of this book is to help with this problem; it does not pretend to substitute for the study of the originals.

It attempts to discuss thoroughly the range of quantitative methods available for particle outlines and silhouettes that have been put forward for various purposes. There may have been good reasons in the past for a method to fail that are not relevant now. For example, the manual use of microscopes was a disadvantage; it was not only slow, but it was also 'unpleasant.'

Rose (1953) wrote:

'It is probably a not unfair summing-up to state that the microscope as a means of powder size analysis should be looked upon as a last resource and is not suitable for routine analysis.'

In the same book Rose also wrote:

'It is also of interest to notice that the accuracy of the observations of a worker are influenced by factors such as fatigue, boredom and ill-health and also by the degree of unconscious mental revolt against visual observation under unnatural conditions.'

The consequences were considerable, but now there are available computer-controlled, automated and semiautomated optical microscopes, user friendly scanning electron microscopes with built-in software-driven image analysis, computer-accessed image recorders, and stand-alone image-analysis equipment. Money can replace man-hours and tedium and provide, very often, improved precision with high rates of working. The conclusion of a recent paper by Jarvis (1988) is pertinent:

'The concept of a low cost laboratory image analyser is no longer the subject of speculation. Limitations of early systems have been surpassed by high performance video and microcomputer technology. This discussion is itself evidence of the rapidity with which technological speculation has become reality. Commercial image analysers based on the current generation of microcomputer hardware now provide a valuable and affordable laboratory facility. Successful development of a task specific image analysis system is dependent on matching of hardware performance to the application, designing software to take full advantage of the facilities of the hardware and paying adequate attention to the influence of video technology on analytical errors. The future of image analysis for microscopy will be inevitably shaped by micro-electronic developments but the degree of success will depend on appropriate software design.'

Perry et al. (1978) have no doubts about the usefulness of optical microscopy:

'With the advent of modern image analysing devices and semi-automatic sizing aids, the microscope has become much more competitive as a routine particle sizing tool.'

Elsewhere, these authors write:

'In some instances the microscope has been able to provide independent evidence, not accessible by other techniques, and has been used as an arbiter. Whilst we do not recommend the use of the microscope to the exclusion of other techniques, we feel that it has an invaluable role to play in particulate characterization of all kinds.'

It must be said that reference to some of the recent literature in powder technology makes one realise that the kind of list of users of outline shape given earlier in this Introduction seems to have been forgotten, even though an accurate list would now be longer still. Since it would not be much of an exaggeration to say that the successful use of outline characterisation in real applications has been in fields other than powder technology, this literature survey has to be quite wide-ranging. An exception to this neglect of other participants in the struggle to quantify shape was a paper by Rumpf and his colleagues at Karlsruhe (Pahl et al., 1973). The 120 references to work published in French, German and English include examples from other disciplines, but such a scope is rare. This may be thought to be an exaggeration, but there is historical evidence of insularity in the science and technology of powders in the past. Just two relevant examples are outlined here, but others will be found in the main text.

The first example is concerned with powder sampling. In a paper published in 1970, Allen & Khan were critical of the poor sampling techniques of powder technologists. Sand and sugar were used to compare cone and quartering, scoop sampling, table sampling, chute splitter and spinning riffler. The authors conclude that the last two are the only techniques worth considering, with the spinning riffler so much the better that it is the only method to use. No mention is made of the similar study by Kaye published in 1962 , which used 125μm bronze spheres (about the same size as the materials used by Allen & Khan) and which showed that the spinning riffler was by far the best sampling device to use. However, Kaye makes no mention of the facts that the chute riffler was compared with what was effectively a spinning riffler by Wentworth et al. and the results showing the great superiority of the latter were published in detail in 1934.

The second example is directly concerned with particle shape. In a eulogy, Freshwater (1987) rightly praises Heywood as one of the founders of 'Powder Technology'. In one section Freshwater explains that Heywood's solution to the problem of particle shape 'is as elegant as his road to it was typical. Seated on a shingle bank at the beach one day during a family holiday, he was struck by the idea of describing the shape of the pebbles forming the strand. He realized that shape has two parts: first there is the form, which is a general property of a whole class of objects, and then there is the relative proportions, which serve to distinguish objects in a particular class from one another.' This is correct, but there is no suggestion that others might have been interested in the shape of pebbles; geologists come immediately to mind. There are no references to work in any other discipline in the two papers by Heywood published in 1937 and 1938. In fact, of course, a lot of thought had been given to quantifying the shape of pebbles and useful ideas had been published since a paper by Wentworth (1919). These publications included the papers by Wadell (1932, 1933, 1935), which were probably the first to distinguish between shape and roundness and are still influential today.

In conclusion to this Introduction, note that there is no Chapter on commercial equipment although some specific apparatus is mentioned in the text. If there were such a Chapter, it could be no more than a guide to the types of facility available at the time of writing. This would be out of date even in the time from writing to publication. There is no alternative but to scan the advertisements, to attend displays at exhibitions, to request demonstrations and to confer with current users of relevant equipment. The last approach is strongly recommended and many users can be traced through learned journals. The tracing process may be speeded up by the use of abstracting publications such as 'Particle Characterization Abstracts' (University of Technology, Loughborough, UK).

2 Sampling and Particle Presentation

It is assumed in this Chapter that sampling is being done from a bulk powder, but, of course, the particle whose outlines are to be studied may have come from other sources. For example, the sampling of airborne particles is a specialised activity with its own distinctive problems and a literature of its own. Two useful texts, which can also help with details of microscopy, are those of Cadle (1975) and of Murphy (1984).

It has been said that 'sampling down to 0.01g, a quantity often required in characterization studies, is the most difficult step in a total sampling' (Kaye, 1981). It is certainly hard to obtain and hard to spread the last few thousand particles onto microscope slides in a random manner; careful statistical analysis has made this clear, even with relatively easily handled material (Hawkins & Davies, 1984). It is worth noting, however, that when individual particles are examined a large number of them may be needed for adequate representation of the population. This is easily overlooked, but does, in fact, assist with the last stage of sampling. In later Chapters, it will be noted that, in some cases, investigators kept their particle numbers down because they were working with powders in which the variety of outline shapes was or was assumed to be limited. It would be unwise to assume this to be the general case without careful exploration.

Sampling for microscope examination is only marginally more difficult to achieve properly than any other stage in the sample dividing of a bulk powder. This is because the aliquots are going to be assumed to be representative of the bulk and, in principle, every particle should have an equal chance of being selected for examination at the end of the sampling procedures. In this book, 'sampling' is used as a synonym for 'statistical sampling' or 'probability sampling'.

In an otherwise valuable article on particle characterisation (Lieberman,1984), it is suggested in a section on sampling:

'Non representative samples can be the result of aggregation during storage or fine percolation during transport, as well as the more frequently blamed inadequate procedures.'

Any procedure that does not provide representative samples, even from heterogeneous material, must be considered inadequate. All samples are subject to statistical variation, but experimental error and bias must be minimised. Each division stage in the process of reducing the bulk must stand up to scrutiny.

Industrial-scale sampling of bulk powders is so many-sided and has to deal with such quantities that, although relevant to this book, it is not an appropriate topic for discussion within it. The principles that will be outlined for laboratory sampling, however, are applicable on the large scale and are clearly presented by Allen (1990). There is also available a particular series of papers and books that can be consulted both for their own value and for a convenient entry into the pertinent literature (Gy, 1967, 1976, 1982, 1983, 1986; Gy & Marin, 1978). The problems of sampling at this level are considerable; the difficulties still warrant a seminar devoted to them (I.Mech.E.,1991). The essential feature of a sampling procedure must be that the material be sampled when moving and not after it has been allowed to heap up. It is suggested in BS 3406 (1961) that a satisfactory method of sampling a group of containers (bags, kegs) is to empty a random selection at a uniform rate onto a short travelling belt conveyor and to sample from the stream of material falling from the end of the conveyor. Segregation is the problem with heaps and containers, but it can also occur, for example, on conveyor belts when transfer takes place from one belt to another. It is this kind of behaviour, in addition to the problems of bulk and inaccessibility, that gives rise to problems in industrial sampling. Stentiford (1979), in a concise and illustrative paper on the sampling of bulk materials in granular form such as ball clays and china clays, makes a telling comment:

'We are more and more surrounded by infallible black boxes. The box itself may be infallible but infallibility begins with the specimen with which the box is presented. The specimen may be one gram taken from a hundred tonnes or more. We place a lot of reliance on that one gram.'

When a sample comes into the laboratory it ought to be representative and should be treated as such. A quotation from a paper published 21 years ago (Allen & Khan,1970) is still, unfortunately, apposite:

'Most analysts recognise that sample selection procedure is of paramount importance but this is not in general true of the powder technologist. He may accept that when powder is poured into a heap, size separation occurs; that when a container of powder is vibrated the fine particles percolate through the coarse and that when powder flows from a hopper it may have a tendency to unmix; all too often he is prepared to scoop off a small sample and attempt to predict the properties of the bulk from the examination of this sample.'

The handling of the laboratory sample both before and after receipt may be summarised as follows:

(1) There must be no alteration to the sample by, for example, environmental attack, reaction, abrasion or aggregation. In practice this entails protection, gentle handling and prompt attention.
(2) The whole sample must be used in any division process to avoid errors, particularly those due to segregation. This need is easily overlooked and may require more than one set of dividing equipment because of the difficulties of scale.
(3) The division process must be a random one to ensure that every particle has an equal chance of selection.

It will be seen later that it is not enough to assume (a) that a particular dividing procedure gives a random performance nor (b) that a random performance with one powder ensures a random performance with all powders. It is necessary to confirm the situation in every case (Hawkins,1988). This confirmation may require statistical analysis. As soon as statistical analysis is invoked into an experimental procedure there follows a need for many measurements to establish significance. If it is felt that such a detailed analysis is unnecessary, it should be kept in mind that there is a great deal of labour to follow with the measurements on the particles chosen for characterisation . Sensible shape measurement carried out with care to achieve a representative set of values is a slow, painstaking and often tedious task. It should not be undertaken lightly and certainly not on an unreliable sample. In addition to reading frequently cited texts such as that by Allen already mentioned, reference to a particular British Standard (BS 3406, Part I, 1986) is advisable and almost essential before starting to sample powders. This Standard lists and describes the methods available and the actions and precautions to be taken. Experimental results are included; these results are useful for comparison with 'in house' results.

Cohesive powders are stable and relatively easy to sample; the necessity is to obtain plenty of small samples chosen at random to cope with any inherent heterogeneity in the material. Such heterogeneity is quite likely in a cohesive powder because of mixing difficulties. One suitable way to sample is to spread the powder out as a thin layer on a rectangular grid. This permits random numbers to be used for the sampling coordinates in two dimensions, the whole depth of powder being taken at each sampling point. If a cohesive powder is sampled as a heap, there is the difficulty of getting at material in the heart of the heap.

It is free-flowing, segregating powders that are difficult to sample reliably. Like several techniques used in powder work, sampling apparatus has come into the powder laboratory from soil science, but different industries sometimes have different methods. In a useful section of his book on working with particulates, Beddow (1980) lists ASTM-published standards on the sampling of some two dozen particulate materials.

The most common sampler is the chute divider. These dividers come in various sizes to deal quickly with different quantities, but all have a riffle top with three boxes. One box is loaded with the powder to be divided and the material is poured from it into the riffle top. Two sets of chutes divide the powder into two lots. One lot is discarded and the other put through the divider again and so on. This type of sampler has been in use for 60 years or so, and the best way to use it has been the subject of study for as long (Otto, 1937). It is clear that the number of chutes available should affect the efficiency of the riffler. This was investigated by Kaye (1964) among others, and he found that, indeed, the number of particle diameters spanned by the chutes is an important variable. It will be realised that great care is necessary to avoid segregation and thus false sampling occurring during the operations of loading and pouring. With some mixtures, it is impossible to obtain random behaviour.

The preferred method of sampling is certainly by spinning riffler. The superiority of a rotary splitter over a chute splitter was demonstrated nearly 60 years ago (Wentworth et al., 1934). A more recent example revealed significant bias when a Jones chute riffle was used to sample magnesium chips (Montgomery, 1968). A revolving riffle showed no bias.

Although there are differences of detail in spinning rifflers from different suppliers, the instruments divide into types all relying on the same principle. This principle should be well understood if satisfactory use is to be made of a spinning riffler. It is perhaps best appreciated by considering a simple conceptual experiment.

If a quantity of powder that segregates owing to a wide size distribution, say, be poured onto a tray as a conical heap there will be a central core richer in the finer grains surrounded by an outer depleted in the finer with, around the base, coarser material (Williams, 1968/69). If a scoop be used to take a sample of such a heap, it is unlikely to be representative of the whole. However, if a spatula with a small spoon end be used to take aliquots and each aliquot be placed in one of, say, ten receptacles in turn or, perhaps better, each aliquot be allocated to a particular receptacle by a table of random numbers, then, when all the heap of powder has been so dealt with the material in each of the ten receptacles will be reasonably representative and, if the spoon end is small, well representative. This is clear because, in the limit, individual grains could be picked up and put in the receptacles in a truly random manner. Although, in practice, this concept works, the procedure is slow and tedious; the automated version of the method in the form of the spinning riffler is extremely successful.

The small aliquots are obtained in receptacles mounted on a rotating frame under a feed of the powder, as in one version shown in Fig. 2.1.

The feed mechanism may be some form of funnel or, more often, and generally to be preferred, a vibrating feeder where the particles move down the vibrating chute by shuffling and hopping. The feed rate on this mechanism can be closely controlled by adjusting the alternating current through the coils of the vibrator.

From what has been said above, particularly as there can be segregation with time as the sample feeds down the chute, the whole sample of powder must be put through the spinning riffler and the total number of presentations of the collectors to the flowing stream should be high. There are at least three important papers about the theory and use of the spinning riffler, but two used a fixed revolution rate of the collector frame (Allen & Khan, 1970; Charlier & Goosens, 1970/71). In a detailed account of mathematical modelling of the operation of spinning rifflers,

Fig. 2.1 A spinning riffler for small quantities of powder; this example has glass sample tubes (Quantachrome Corporation, Syosset, N.Y./Techmation Ltd., Edgware, Middlesex)

Hatton (1978) implies that the previous writers assumed that the effect of increasing the revolution rate of the collector frame is indistinguishable from that obtained by a commensurate reduction in feed rate. This is unlikely because Charlier & Goosens wrote:

'The spinning riffler, whose operation principle has been fully described by Allen (1970), presents two variable operating parameters, namely the revolution rate and the feeding rate of the solids. For practical reasons, only the second parameter has been considered here.'

In his paper Hatton shows that the two procedures, of increasing the revolution rate and reducing the feed rate, are not necessarily equivalent. 'In order to attain a given number of presentations it is preferable to reduce the mass flow rate as much as practicable rather than to only increase the rotation rate.' This is fortunate, as, if the rotation rate is high, there is a risk of air currents affecting the fall of fine material to such an extent that it is lost from the spinning riffler

If the feed rate in a spinning riffler is properly selected with respect to the time availability of any particular cup, good random performance is obtained. A strongly segregating powder may be made up from glass ballotini of different sizes, but even with such material the spinning riffler is successful. The results obtained from two mixtures made up by weight from three sizes of glass ballotini are given in Figs. 2.2.

The mix values found in each of the 16 tubes of the spinning riffler were very close to the made-up value in both cases. This is a simple illustration of one of the principal features of the spinning riffler: it samples unmixed just as efficiently as mixed powders.

In the most commonly used spinning riffler, the ratio (total time of powder flow)/(time of rotation) should be large, of the order of 100 or more. The desirable ratio can be calculated, but, as Charlier & Goosens state, 'it seems reasonable to assume that the splitting of a batch in representative samples will be easier when the degree of homogenization of the mixture is higher. For instance, only one complete revolution of the splitter is necessary for a perfectly homogenized mixture. The assumption introduces then a second parameter which will be referred to as the solids feeding model.' The authors proceed to show how to deal with three different feeding models.

With a vibratory feeder, there can be segregation across the width of the chute. The whole fall must be received by the receptacles, but there is no need for a complete set of receptacles; true random sampling can replace random dividing. With the material falling as a curtain from the vibrating feeder at a suitable rate, just two large receptacles rotating on the carrier will collect random samples. The unwanted material will collect up in the carrier, and the two samples can be combined as the riffled sample.

When a full set of sampling containers is used and not all of the contents is required, opposing pairs should be mixed to give a riffled sample. The spinning

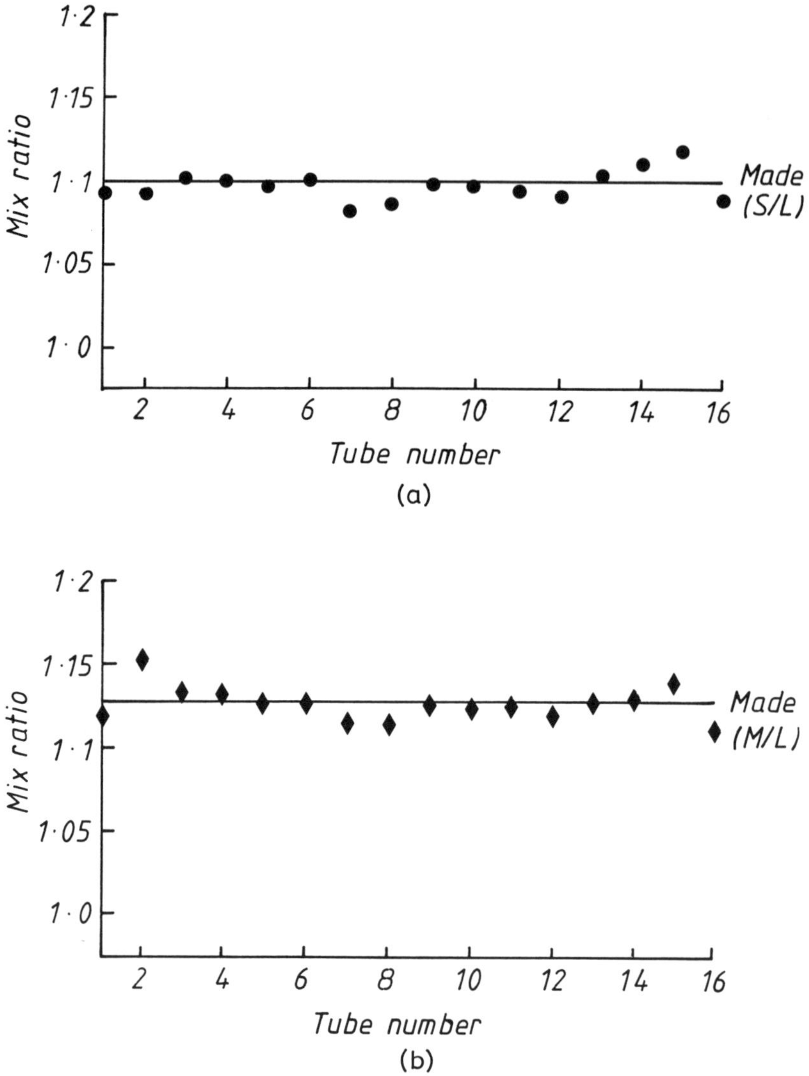

Fig. 2.2 The ratio of (a) (+63-125μm)/(+300-600μm)
(b) (+180-250μm)/(+300-600μm)
glass ballotini found by sieving the subsamples of made-up mixtures put through a spinning riffler
(Reproduced from A.E. Hawkins Part. Part. Syst. Charact., 1988, **5**, pp23-28, by permission of VCH Verlagsgesellschaft mbH)

riffler should be made with knife-edges between receptacles so that a particle goes one way or the other and there is no build-up between receptacles. A particular arrangement of spinning riffler, very useful for small quantities of powder, does have a limitation in this respect and attention needs to be paid to this difficulty when using the device.

This particular sampler is shown in Fig. 2.3.

The triangular stainless-steel receptacles are convenient to use with small quantities of fine powder since they can be cleared out with a good-quality artists' brush. Retention of finer particles by adhesion to the glass sample tubes can be a problem with some spinning rifflers and could lead to bias.

There is a group of dividers in which powder is fed to a rotating chute that distributes the powder to stationary cups. This rotary sampler is faster to use than the usual spinning riffler and is more easily totally enclosed to avoid contamination. Only the feed rate to the rotating chute can be controlled, and care is needed if reliable sampling is to be achieved.

Unlike the other types of sample divider available, the spinning riffler is very nearly operator-independent provided, as mentioned earlier, the operating ratio is kept at a suitable value. As a general rule, the larger the number of receptacles rotating under the curtain of falling particles at a particular rotation speed, the easier it is to avoid any operator effect.

Special forms of chute divider have been devised to produce the very small quantities of powder needed for individual microscope slides. These do not seem to be available commercially, but are designed and made by individuals. An example of such a microsample splitter that assists the free flow of grains is milled from solid brass (Humphries, 1961). A diagram of the device is reproduced in Fig. 2.4.

When the particles have been distributed over a set of microscope slides, then, if possible, every grain should be measured. If this is not practicable, sampling of the grains on the microscope slides must be considered. If not all the slides are to be examined, those that are should be selected in a random manner before they are examined. If not all the grains are to be measured on a slide, they too should be selected in a random manner. Sampling on a slide is no different in principle from sampling over an area in, say, geography or geology. For random sampling, two random numbers are taken, one representing the E-W direction from some origin, and the other representing the N-S distance from that origin. Correct application of this technique may not be easy to achieve if the microscope is not equipped with a mechanical stage and verniers since coordinates cannot then be selected by random numbers. One solution is not to use ordinary microscope slides, but the far more costly England Finder Graticule Slides. These are divided up and numbered as shown in Fig. 2.5. reproduced from a laboratory supplier's catalogue.

Each square contains a centre ring bearing a reference letter and number and is further subdivided.

A large amount of statistical theory arose from the analysis by R. A. Fisher and others of agricultural studies at Rothamsted, so that sampling an area has been much considered. For a detailed examination of the alternative strategies available,

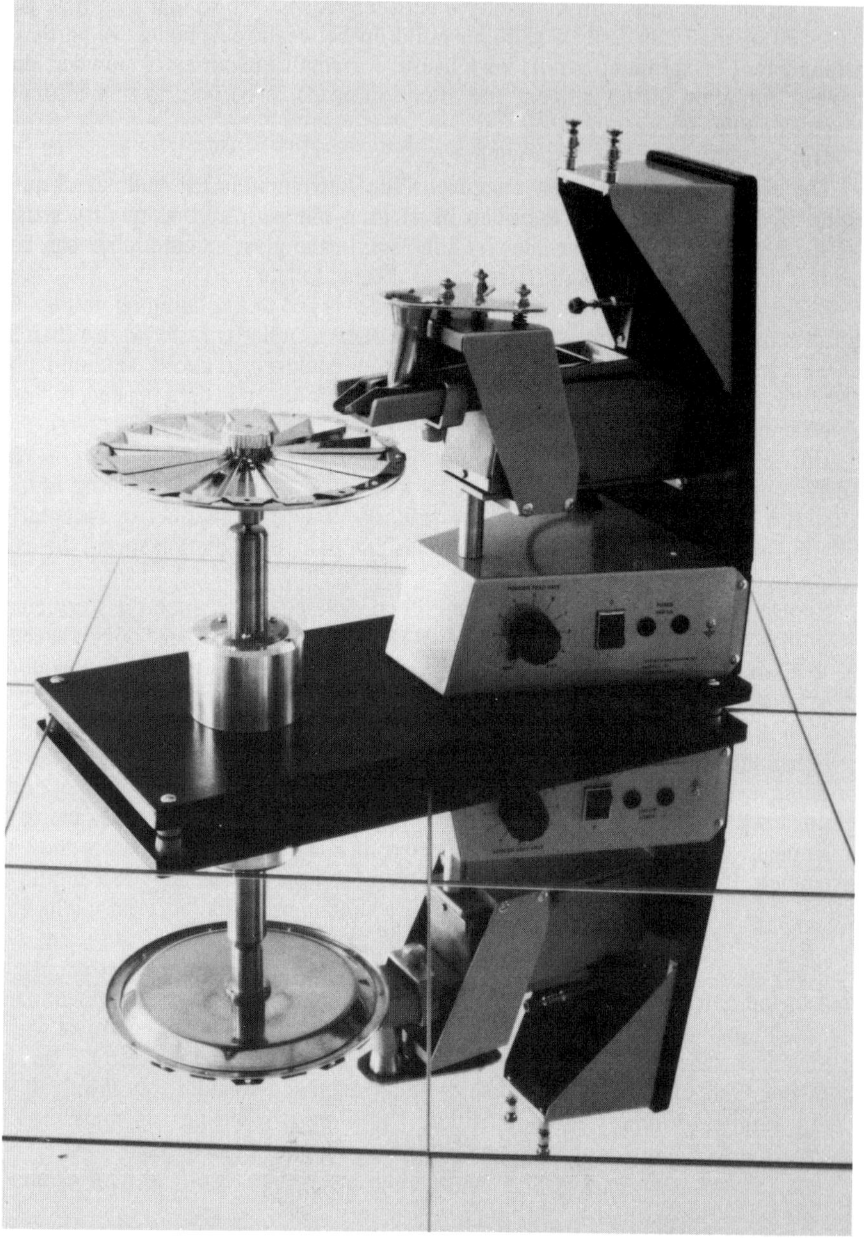

Fig. 2.3 A spinning riffler for small quantities of powder; this example has stainless-steel collectors. To improve the feeding of cohesive powders, the riffler is equipped with a stainless-steel hopper attached to the vibratory feeder chute (Microscal Ltd., London)

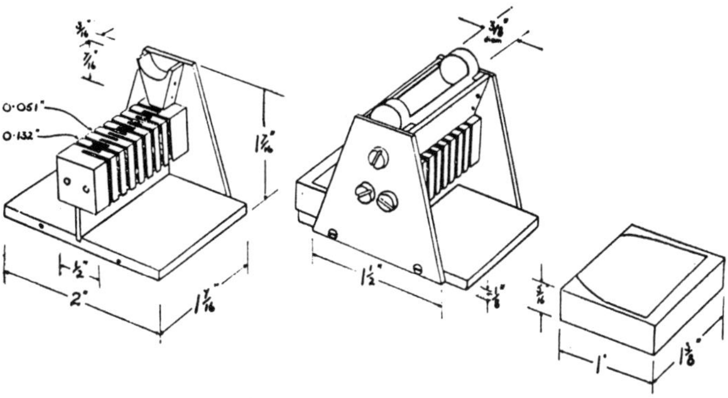

Fig. 2.4 A microsplitter for use with microscope slides (Reproduced from D.W. Humphries, *J.sedim Petrol.*, 1961, **31** pp471-473, by permission of the Society for Sedimentary Geology)

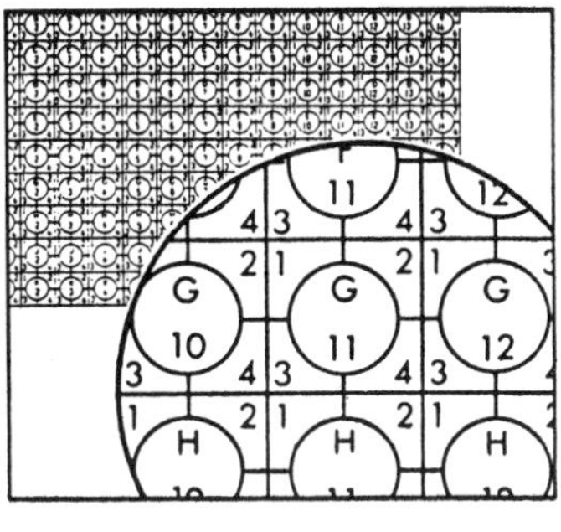

Fig. 2.5 Diagram of the England Finder Graticule slide. The rows are lettered, the columns numbered. (Reproduced by permission of Fisons Scientific Equipment, Loughborough, Leics.)

reference may be made to a suitable text such as that by Cochran (1977), but simple random sampling will generally be found satisfactory. A paper describing an automatic sampling stage microscope (Weibel, 1970) is helpful on the principles of random sampling applied to microscopy and shows when systematic as opposed to simple random sampling is to be preferred.

It was noted above that the sample of particles on the microscope slides should provide adequate representation. This is made more complex by the fact that the outline is only two-dimensional. Although examination through a stereomicroscope can reveal flakes or chunks or spheres that is not the same as their measurement. The presentation to the microscope is very important. In the literature, two views will be found, but quite often the subject is ignored. The two views are (1) put the particles in their most stable position under gravity, particularly when measuring size; (2) present the particles in a random array.

Going back half-a-century, Riley (1941), when reviewing some previous accounts of projection methods for examining microscopic fragments, notes that neither Pentland (1927) nor Cox (1927) specified any definite orientation of the grains, but that Tickell (1931) specified a random orientation. Riley himself used the same presentation as Wadell (1935): maximum cross-sectional area, i.e. the cross-section containing the long and intermediate diameters. At about this time it was stated 'If the grains are to be mounted in the dry state, the final sample may be transferred to a clean microscope slide, and the edges of the slide tapped with a pencil, so that the grains assume their most stable position of rest.' (Krumbein & Pettijohn, 1938).

In describing how grains may be studied and their outlines measured, Cox (1927) says 'the grains may be sprinkled over a lantern slide that has been previously covered with glue or a suitable cement, the slide inserted in a projection camera, and the image obtained on the screen.' Unless surface tension has an influence, this automatically provides a random presentation, although the author does not mention this. In another place, Cox remarks 'when grains have a decidedly tabular form, such as mica particles, it is necessary to imbed the material in some supporting medium and cut a thin section at random in order to insure a true sample of the shapes.'

Tickell (1931) writes:

'It should be borne in mind that both of these roundness numbers are used on the assumption that the grains are not lying on a greatly flattened side, as is the case with mica, etc. If, for example, the grains are predominantly feldspars, and are rounded cleavage flakes, they should be imbedded in balsam and a polished surface made. This surface may be photographed by reflected light and the grains will have random orientations.'

In powder work, the object of study is generally a projection or silhouette of a particle. It is perhaps as well to keep in mind that the outline being studied on the microscope slide may be of a more difficult form, such as a section through some object. This is especially true in the life sciences, geology etc. The section may be natural or from an object cast in, say, epoxy resin. Sectioning is also used in the study of ceramics and in powder metallurgy, where the shape of the spaces between the particles, the pores, can be more important than the shape of the particles themselves. The differences are considerable, both in sampling and in presentation. Because the section may be anywhere through the object, statistical extrapolation is made from two- to three-dimensional space. This is quantitative stereology (Underwood, 1970; Weibel, 1979) and adds a further complication to an already difficult task.

The fact that the shape of a particle outline is rarely some convenient geometrical figure has been a problem in size measurement by microscope for many years, particularly because it led to several definitions of size. This was true in the days when 'diameters' were directly measured with a micrometer eyepiece, and is still true today with very-high-speed automatic image analysis. A relatively modern paper points out errors in the contemporary microscopic measurement of particles. Davies (1962) writes:

'Irregular particles, seen as silhouettes, can be measured in various ways. What is meant by the terms radius, diameter, size, is frequently unspecified and figures derived from such measurements are seldom related to the statistics of shape and then only by factors of doubtful meaning.

Any of the common methods of expressing the sizes of the particles in a preparation can, however, be associated with the average of a series of analogous measurements of a single particle, oriented in different ways; this average has an exact geometrical significance.

It is suggested, therefore, that if the repetition of the measurements be made not at random on a single particle, but once per particle over a number of particles lying randomly, then their average will have the same geometrical significance for the group as the average of the measurements taken on a single particle has for that individual.'

Tilman (1973) found that for quartz the view was the same in all directions. This was on one sand and for 50 grains 'selected at random from a Lake Michigan beach'. This seems a somewhat inadequate base for the inherent belief in several later papers that a representative set of shapes had been obtained. The silhouettes from photographs in Fig. 2.6. show 45μm shape-sorted sand produced by Pang & Ridgway (1980). Samples from the extreme cups are shown with a random presentation, and, clearly, the shapes are different in the two cases.

Almost ten years before the publication of the paper by Tilman, attention had been drawn to the difficulty of quartz particle shape (Lees, 1964). The author suggests that, because of the rarity of the oblate (discoid or bladed) form in quartz grains, it was generally assumed that the flatness ratio, i.e. (shortest dimension)/(intermediate dimension), was close to unity in all specimens. 'Whereas such an assumption may be justified for uncrushed quartz sands, it is not where minerals of different crystal habit are present in significant amount, or where a significant quantity of crushed material is present, *even if this is still all of pure quartz.'*

It is clear that it is unwise to assume anything for any material. Grains should be put on slides that are partly sticky and partly not and some relevant measure of the several to be described in following Chapters be used to establish the need or otherwise of random presentation. Of the many slide coatings that can be used, rubber adhesive diluted with hexane has been found useful, but double-sided adhesive tape is often adequate.

If the particles are in a suspension, there can be problems due to sedimentation within the storage vessels. The powder particles may fall to the bottom of the vessel, form a cake and be impossible to return to their original state. The effect of particle shape on such caking has been investigated for antacid suspensions (Heyd &

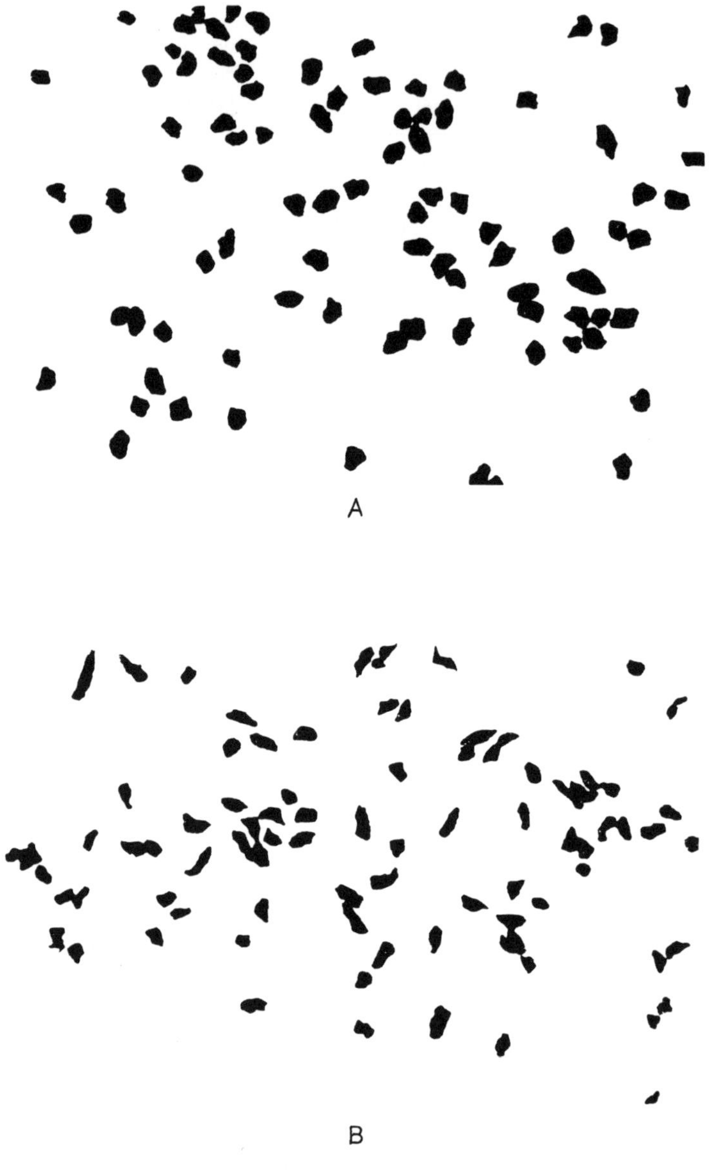

Fig. 2.6 Shape-sorted sand. A = pot 1; B= pot 13

Dhabhar, 1979). This does mean that samples should be dealt with promptly and, if there is no danger of particle break-up, be kept in agitation by, say, putting the storage vessel in an ultrasonic bath.

Great care must be taken in sampling from a suspension. The separation of sizes due to gravity can easily cause a sample to be biased towards either the finer or the denser fraction (Davies,1968/69; Allen & Baudet,1977). There is only one way of guarding against this happening, and that is to take many samples from the suspension and determine whether they are statistically all from the same population. If such a test of a sampling technique is required it might be convenient to use for the purpose size distributions by electrozone measurements. These are both reliable and relatively rapid and should show if derived data are changing as successive samples are included. A useful paper relevant to this topic is that by Polke et al. (1991). Although concerned with dispersions for size measurement, the ideas and findings are applicable to shape characterisation as well.

If a small aliquot from a suspension is spread out over a microscope slide, there are two potential sources of error: (1) aggregates may be broken up by the mechanical action of spreading the sample; (2) larger particles may settle out in the centre of the slide and the fines be taken to the extremities by surface tension. The former can be circumvented by gentle agitation and the latter by examining every particle on the slide. It is this problem that requires random sampling of the slide to be used rather than, for example, sampling along certain horizontal or vertical lines.

It will be realised that particles from a suspension will probably settle into positions stable under gravity. This is true also of an otherwise very serviceable method of obtaining prepared slides using a eutectic (Thaulow & White, 1971/2). The eutectic recommended is of 60 wt.% camphor, 40 wt.% naphthalene which melts at 32° C. The particles are mixed in and the mixture is spread on slides and allowed to set. The material sublimes completely under vacuum avoiding particles being drawn together by surface tension, as can happen when evaporating a suspension.

A common problem when making microscopic measurements on particles on a slide is that of particles touching or overlapping one another. This can be important. Despite manual operation it can be difficult or impossible to make measurements on two touching particles and still more of a problem if one particle is partly overlapping another.

Even if the whole area of a microscope slide is being used for particles, and that is a difficult arrangement to achieve, surprisingly few particles can be sprinkled on a slide before touching and overlapping begin to occur. This can be illustrated by a simple and approximate application of probability theory (Hawkins, 1987).

A microscope slide is first visualised as covered with squares of side equal to the maximum particle dimension. If there are Q squares on the slide and the first particle is somewhere upon it, the probability that another particle does not fall on the same square, nor upon an adjacent square, is $(1\text{-}9/Q)$.There are now two particles on the slide, and the probability of an acceptable fall by a third particle is given by $(1\text{-}18/Q)$ and, for a fourth particle, by $(1\text{-}27/Q)$, and so on.

The probability of one unacceptable fall (i.e. just two particles touching or overlapping) when n particles are dropped is given by P_S, where:

$$P_S = 1 - \left[(1-9/Q)(1-18/Q)(1-27/Q) \ldots (1-9(n-1))/Q \right]$$

This expression approximates to:

$P_S = 9(n^2-1)/2Q$ because $9/Q$ is small.

The use of regular hexagons instead of squares improves the realism and leads to the equivalent probability $P_H = 7(n^2-1)/2Q$.

The number of particles at which the probability of a touch becomes unity (i.e. a certainty) is quite small, even though it is assumed in the calculation that the whole of the slide to the edges is being used. For example, the values are about 50 for 500μm particles and 240 for 100μm particles on a standard 76mm x 24mm slide. This has been tested experimentally and has been found to be optimistic; fewer particles than are indicated by calculation should be assumed as safe to use because, as already mentioned, it is difficult to make use of the whole of a microscope slide.

The apparatus shown in the photograph of Fig. 2.7. is a 'moving pavement' version of a spinning riffler.

A vibratory feeder drops the particles onto slides placed on numbered carriers fixed to a driven rubber belt. This can be used equally well with plain or coated slides. There is a limit to the speed at which the slides may be moved, but control of both the linear speed of the slides and of the particle feed rate was found necessary (Hawkins & Davies,1984).

The performance of the unit was subjected to stringent statistical investigation and was found wanting at the original fixed speed of 44 mm/s, despite favourable appearances of the particles on the slides. The original paper contains important detail, but some condensed data will illustrate the key points.

If there is an equal chance of a given particle falling on any of the slides, the final number of particles on each slide should follow a Poisson distribution (Student, 1907; Lancaster, 1970). Moreover if each slide contains a random number of particles from the parent population, the number of particles in any size range should also follow that distribution. The testing of the distribution for absence of bias is essential.

The mean value, $\bar{x}$, of a Poisson distribution is equal to its variance, so the standard deviation σ is related by $\sigma/\sqrt{\bar{x}} = 1$

The particles should be spread evenly over all the 34 slides in the apparatus. If this is done perfectly, one may expect the same number of particles on each slide. A measure of departure from this is given by:

$$\sum_{i=1}^{n} \frac{(O_i-E_i)^2}{E_i} = \frac{(n-1)\sigma^2}{\bar{x}}$$

here O_i = observed number on slide i
E_i = expected number on slide i
= $\bar{x}$ for all i.

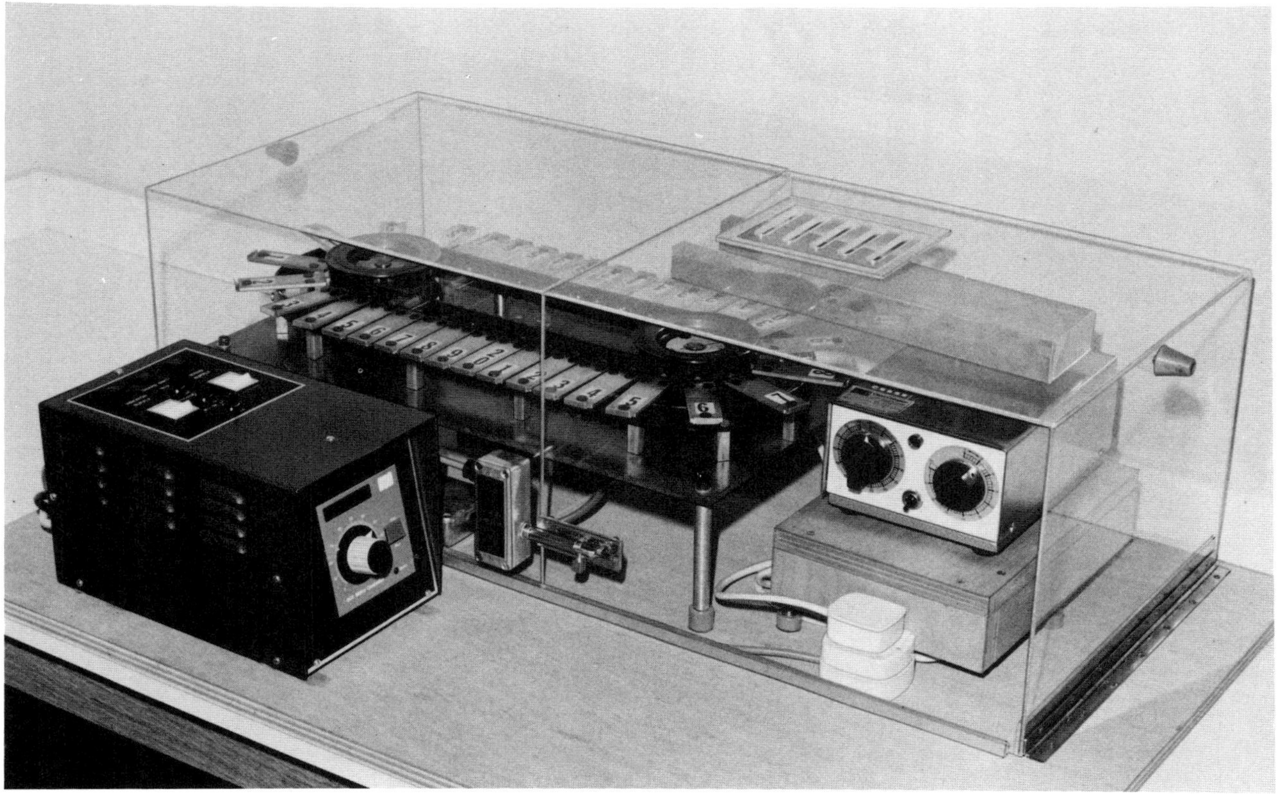

Fig. 2.7 A form of spinning riffler for microscope slides

This statistic has a chi-square distribution with (n-1) degrees of freedom and thus permits the computation of the chance that the ratio $\sigma/\sqrt{\bar{x}}$ will be exceeded in a Poisson distribution. For 34 slides, if the distribution is truly random, the value of 1.39 will be expected to be exceeded on 1 in 1000 occasions.

The values for the projected area diameters of a sample of sand spread over the 34 slides and for a sample of spray-dried powder similarly spread are given in Table 2.1.

Table 2.1. Values of the test coefficient ($\sigma/\sqrt{\bar{x}}$) for a Poisson distribution of projected area diameters for sand and a spray-dried powder.

	Sand			Spray-dried powder	
µm	$\bar{x}$	$\sigma/\sqrt{\bar{x}}$	µm	$\bar{x}$	$\sigma/\sqrt{\bar{x}}$
140-180	32	1.7	53-106	72	2.5
180-200	48	1.7	106-150	204	2.3
200-220	65	2.0	150-212	313	5.5
220-240	81	1.9	212-300	299	5.9
240-260	65	2.0	300-424	166	1.8
260-280	32	1.9	424-600	62	1.3
280-300	11	1.5	600-2400	14	1.5
300-420	7	1.3			

This unsatisfactory performance, with the test ratio generally much greater than the desired value of 1.4, was identified with particles coming from the feeder in groups; surges could be detected by autocorrelation of the results from adjacent slides.

The feed rate was already low so a variable speed drive was fitted to the 'moving pavement' arrangement. A speed as high as 200 mm/s was found to be possible and could be adjusted to suit the powder concerned. With this facility and the feeder adjusted to have the particles hopping about on the end of the chute for several seconds before dropping off, a good performance was obtained.

The phrase 'the most difficult step' quoted at the beginning of this Chapter should, perhaps, have read 'the most revealing step'. Poor sampling of particles onto a microscope slide is soon revealed; poor sampling of a quantity of powder is more difficult to identify. Fortunately, the essentials of dry-powder sampling are straightforward to work at to achieve:

1. A sampler should be used whose results are independent of the mixed state of the powder.
2. The full quantity of a powder stream should be taken for short increments of time, rather than part of the stream for the whole time.
3. The aliquots arising from the short increments of time should be small, so that the total number of aliquots is very large.

The purpose for which the particles are being examined must be kept in mind, because this will determine whether the particles should be presented to the microscope in a random manner or ordered in a state of stability under gravity.

The need for careful consideration of the purpose of the measurement of particle- outline shape will become clearer in the following Chapters. These Chapters progress from basic measurements to measurements so detailed that the meaning of the data is quite awkward to extract. In a more general context, this situation was summarised over 50 years ago (Krumbein & Pettijohn, 1938):

'Many workers in sedimentary petrology question the value of mathematical analysis, owing to the large number of variables and errors which are involved in the simplest situation. In general, this may be granted in the present stage of development of the science, but it is equally true that an approximate determination of the nature of the fundamental relationship may point the way for more rigorous studies of problems and eventually establish underlying principles of universal application.'

More recently Tsubaki & Jimbo (1979) stated:

'From the point of view of the behaviour of powders, it is necessary to express statistically the morphological characteristics of particles which affect behaviour, rather than to reproduce exactly the particle shapes.'

This is a rather sweeping statement, because these two aspects of particle-outline shape are not necessarily mutually exclusive. There is also sometimes a need for particle identification alone without reference to the bulk powder. For these reasons, both types of shape characterisation are dealt with in the following Chapters.

3 Qualitative Description

Although it is taken for granted in this book that it is desirable to measure the shape of particle outlines, there has to be a consideration of what is meant by the term; this is not a simple matter. A glance at the 38 silhouette shapes, used by Heywood (1945) to compare methods of measuring the size of particles with a microscope, will illustrate the problem.

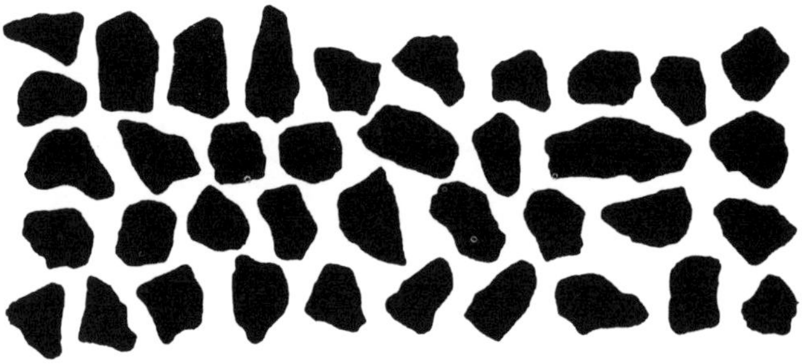

Fig. 3.1 Silhouettes used by Heywood to compare methods of measuring microscopical particles (Reproduced from H. Heywood, *Trans. Inst. Min. Met.*, 1945, **4**, pp391-404, by permission of the Institution of Mining and Metallurgy)

These silhouettes were of quite large particles; similar silhouettes may be produced from small particles with a transmission electron microscope, but, even with these, most of the literature deals with outlines that are basically regular and not, for example, dendritic.

There have been many words written about the meaning to be given to the term shape when applied to road stone, rock chippings, pebbles and, more recently. powder particles and particle outlines. Dictionaries have been cited, but to little purpose, and it is necessary to consider some of the literature to determine the consensus of opinion about shape before embarking on its measurement. It will be found that, as the study of shape has continued over the years since Sorby's scheme of classification (1880), certain terms have been redefined, and care is necessary in their use. In this account, the shape of a powder particle will be taken to include every aspect of external morphology, i.e. the form, the sharpness of any edges or corners and the texture of the surface; this notion is the same in two-dimensional terms except that an outline or silhouette can hardly have a morphology. This concept of shape is illustrated in Fig. 3.2.

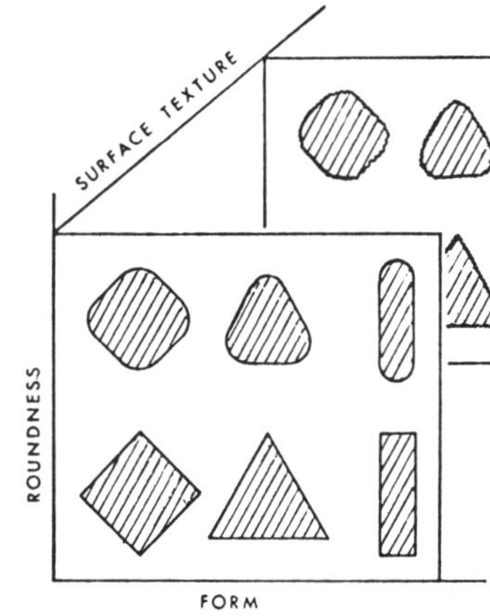

Fig. 3.2 A simplified representation of form, roundness and surface texture to illustrate their independence (Reproduced from P.J. Barrett, *Sedimentology*, 1980, **27,** pp291-303, by permission of Blackwell Scientific Publications Ltd.)

This Figure is a simplified representation of form, roundness and surface texture by three linear dimensions given by Barrett (1980) in an illuminating review of investigations into the shape of rock particles. The independence of the three aspects is clear. Müller (1967) explained form when he wrote that the morphology of a particle is determined by its relative dimensions (e.g. rod-shaped, equant, tabular, blade, spheroid, disc, roller etc.), by its roundness (very angular, well, moderately, or poorly rounded etc.) and by its surface texture. Pahl et al. (1973), in the review on methods for describing particle shapes mentioned earlier, write:

'Four characteristic features which are independent of one another can be quoted for describing the geometry of particles i.e. size, shape, roundness and roughness.' This clear concept was not around from the beginning of particle-shape studies and, as indicated earlier, the development of ideas about shape has been a continuous and is a continuing process.

In parallel with pragmatic studies of shape, there has been the development of a psychophysics of form perception dating back perhaps to an early paper on shape and pattern perception (Attneave & Arnoult, 1956). These particular authors make significant comments:

'Likewise, when some experimental variation of form is found to produce a certain effect in the laboratory, it is necessary that the variable in question be identifiable and measurable outside the laboratory if the results are to be generalized. Unfortunately, however, it is much harder to measure form than to manipulate it.

Relatively few scientists have seriously applied themselves to the problems of analyzing and describing form; these problems seem to have fallen into the cracks between sciences, and no general morphometry has ever developed.'

A suitable entry date into utilitarian research is 1932, because it was then that Wadell called attention to the fact that the use of the word 'shape' required attention. Previous workers, for example Trowbridge & Mortimore (1925), had used a visual method of determining 'roundness' by comparison of the material under question with a more or less arbitrary set of standards, and both Pentland (1927) and Cox (1927), had made measurements and called them 'roundness'. Wadell pointed out that 'roundness' was a matter of the sharpness of the corners and edges of a grain, but that there was another independent variable, 'shape', which had to do with the form of the grain independent of the sharpness of its edges. This concept is illustrated for two-dimensional outlines in Fig. 3.3, but Wadell was, in fact, more often dealing with larger particles on which three-dimensional measurements could be made. This does not matter in this context, but it will be seen in the next Chapter that it had an effect on his work on the quantitative measurement of roundness and shape.

Wadell was interested in sedimentology and used the sphere as a standard of reference because of its behaviour when sinking in a liquid and because a sphere will roll more easily than solids of other shapes.

In the next Chapter, it will be seen that Wadell quantified 'shape' by a mathematical expression to which he gave the name 'sphericity'. This term 'sphericity' is based on the proximity of a particle to a perfect sphere. In two-dimensions, 'circularity' should be based on the proximity of a particle outline to a circle. Unfortunately, this is not always the usage; sometimes the usage is neither sensible nor logical. Wentworth (1933) disagreed with the Wadell terminology, but the latter prevailed.

Marwick (1937) wrote that it is useful to distinguish between two distinct ideas. In the first place, there is the relative roundness or angularity of the material, which it is convenient to term the 'form', and in the second place there is the 'shape' proper, by which is meant the extent to which the particle may be said to be flaky or elongated. This, again, is contrary to the Wadell view, which has prevailed, certainly in geology.

Heywood (1937) wrote that it was essential to distinguish between the shape and the proportions of a particle. Shape, he declared, may be defined by the closeness with which the particle approximates to certain geometrical shapes such as ellipsoids,

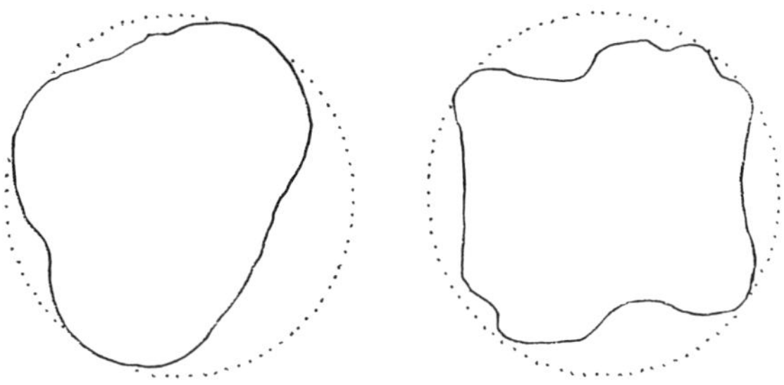

Two outlines of the same sphericity, but different roundness (same outline area; same smallest circumscribing circle; different average radius of curvature of corners)

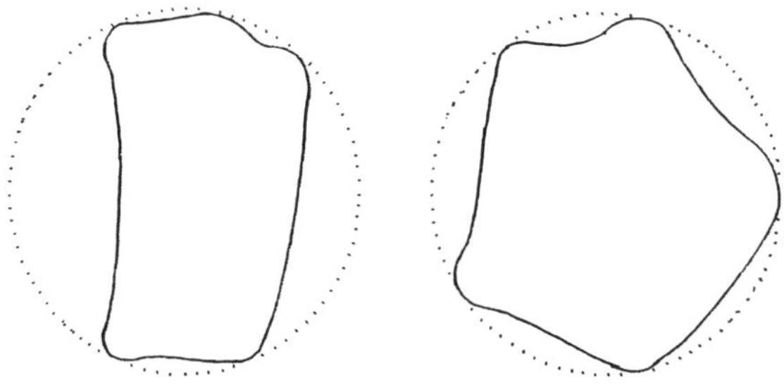

Two outlines of the same roundness, but different sphericity (same average radius of curvature of corners; same smallest circumscribing circle; different outline area)

Fig. 3.3 Illustrations of sphericity and roundness

prisms or tetrahedrons. The proportions of length, breadth and thickness of a particle are factors quite distinct from the shape: thus, either a flat or an equidimensional particle may have any of the geometrical shapes mentioned above.

This comparison of shapes with some geometrical form had been put forward by Wentworth (1936) with words such as prismoidal, bipyramidal, pyramidal, wedge-shaped, parallel tabular etc. to describe the appearance of grains. For viewing from above, Wentworth suggested describing the form of the margin or outline as hexagonal, pentagonal, trapezoidal, oval, rhombic etc. A silhouette or an outline seen via a microscope may reveal little of the material in the third dimension. In the simplest terms, a disc, a vertical right cylinder and a sphere have the same silhouette.

Powers (1953) considered that an ‘impression’ of the roundness of a particle can be obtained when using a microscope by focus adjustment from top to bottom of the particle. An alternative and, if the particle is greater than about 20μm, helpful technique is to use a stereomicroscope. The combined view, from the two optical systems in this type of instrument, provides a very clear idea of the particle in depth. Some stereomicroscopes permit photography with viewing in the same format, using transparencies in a special projector, or prints in a viewer. It is, apparently, feasible to modify a one-objective binocular microscope for stereoscopic viewing, and, in this case, it is available to high powers (Sroczynski, 1990). Another recent development is the introduction of a research-class stereomicroscope (Wild M10), which is said to provide a resolving power previously only obtained from compound microscopes (Leica UK Ltd., Milton Keynes, Bucks.)

Wadell used a term ‘image’ to sum ‘sphericity’ and ‘roundness’. This combination does not seem to have been used much, but an interesting echo of it is in a paper by Hausner (1966), in which he says that it is helpful if shape descriptors permit an image of the actual shape of an individual particle to be visualised. This is a point not often made in the literature, but Medalia (1970/71) says his numerical shape factors are not intended to replace the usual qualitative descriptions of shape, such as ‘dendritic’, ‘acicular’, etc., and should, preferably be used in conjunction with such descriptions.

In this connection, words such as ‘rounded’, ‘well rounded’, ‘angular’, ‘subangular’ may be used, but no two workers will select the same division point between two classes. The use of model shapes to illustrate the words can help with this difficulty and, in the next Chapter, it will be shown how this has been combined with numerical values. A set of shapes modelled in clay published by Powers (1953) is reproduced in Fig. 3.4.

Other authors have used and published silhouettes and yet others photographs of ground stones, but these are all perhaps more easily used when dealing with material that is large enough to be viewed without a microscope.

There is a British Standard 2955 (1965) that includes terms like:

acicular:	needle-shaped
angular:	sharp-edged of roughly polyhedral shape
dendritic:	branched crystalline shape

fibrous : thread-like
granular: approximately equidimensional but irregular
lamellar: plate like

and so on with crystalline, flaky, irregular, nodular, spherical etc.

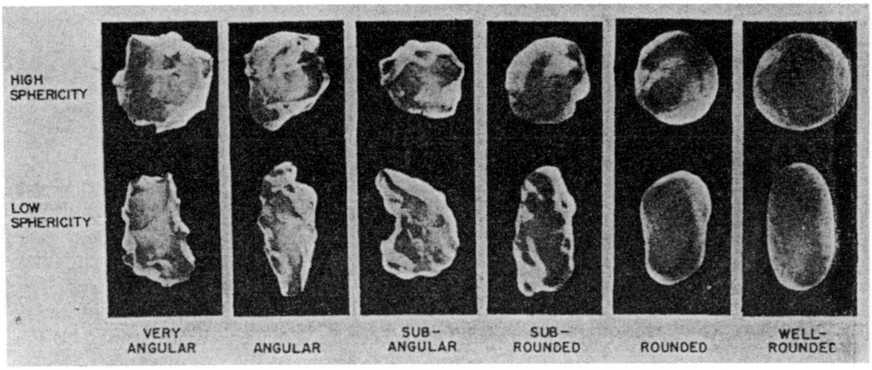

Fig. 3.4 A roundness scale (Reproduced from M.C. Powers, *J. Sedim., Petrol.*, 1953, **23,** pp117-119, by permission of the Society for Sedimentary Geology)

Examination can be all that is needed if the sampling is adequate. For example, if some particular particle shape must represent less than some specified proportion of a bulk powder. It may also happen that examination is the only approach possible; dendritic shapes, for example, present a challenge to characterisation difficult to meet.

Such an examination and photomicrography could be enough if a record of past successful material is needed in an archive. It is not sufficient to be concerned about shape when something goes wrong. It is necessary to have ongoing records; memory is unreliable and, possibly, prejudiced. A photographic record used to be the only option, apart from drawings made through a camera lucida, but both suffer the same demerit: rapid, accurate recall is quite difficult to achieve. Today, the archiving of microscope images has been much improved by the introduction of optical video technology. The video disc can be accessed via a hand controller to locate individual frames on the disc that contain the individual micrographs, but the real power comes from computer drive of the player (Newman, 1989).

Less sophisticated, but often adequate and easily computer-stored and -accessed, is a polar diagram from any point within the outline, as in Fig. 3.5.

The values of the polar radii taken at equal, small angles are a sensible recording system on computer disc, and the polygon approximation of the outline is easily reproduced by computer and printer. If small angles are used, the polygon reproduced can be quite adequate; in Fig. 3.5, it is seen that even 10° angles provide a

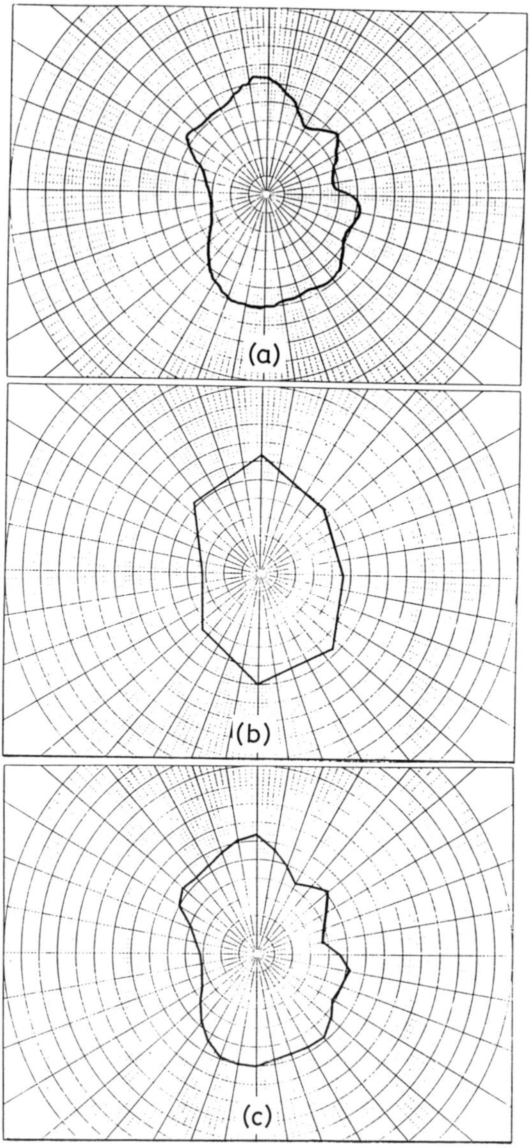

Fig.3.5 Polygons representing a particle outline (a) original, (b) 8 point, (c) 36 point

record that could be adequate although detail is lost and curves are sharpened.

It will be obvious that every polar radius must be single-valued. A re-entrant situation, as in Fig. 3.6, is not permissible.

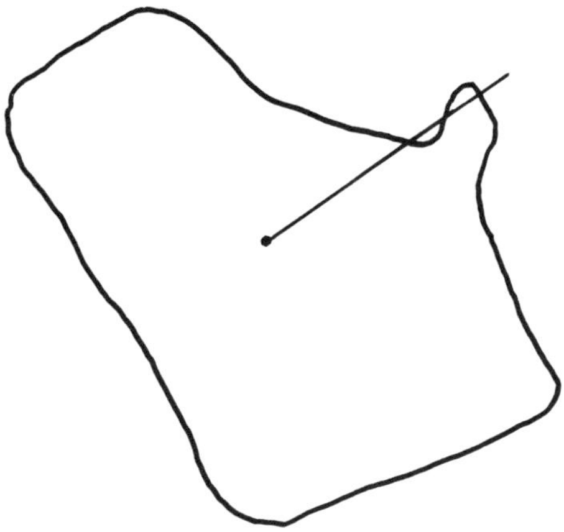

Fig. 3.6 An example of a multivalued radius vector

This feature is not a common occurrence (Walton, 1948); but, if it does occur, it may be possible to overcome the problem by changing the chosen position of the origin.

The temptation to consider the distribution of radii lengths in a polar diagram as useful information alone should be resisted because sequence must not be omitted. The complication is illustrated in Fig 3.7. for the simple example of a 45° plot with four equal long and four equal short radii.

Two very different outlines are produced when the sequence is changed.

Polar diagrams are a positive step towards the quantitative description of particle-outline shape, but a headlong rush into measurement of particle shape should be avoided. Intellectual effort to create mathematical systems may be rewarded by publication in the literature, but the labour involved in making the measurements to generate the data for their manipulation may not be adequately rewarded by the resulting information for use by the experimentalist.

An example where quantitative information was not needed is in the paper by Heyd & Dhabhar (1979). The authors were concerned at the different caking behaviour of coarse granulated antacid suspensions containing two calcium carbonates with nearly identical physical properties except for particle shape. One brand of calcium carbonate had 'asymmetrical needle particles'; the other had 'symmetrical

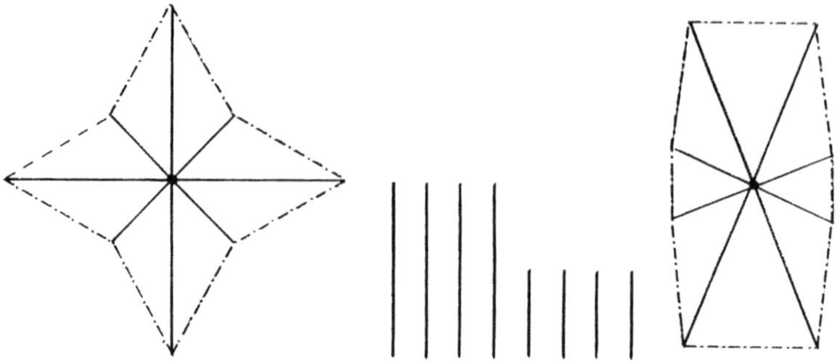

Fig. 3.7 Two different outlines that may be generated from eight radii at 45° intervals by altering the sequence

barrel-shaped particles'. The needle-shaped material produced unacceptable caking behaviour.

An example outside powder technology, where categorisation by observation through a microscope was successful, is instructive. The two dimensional outlines of polymorphonuclear leucocytes were assigned into four categories: spherical; slightly polar; moderately polar; strongly polar (McCarthy et al., 1990). This provided a rapid and valuable diagnostic procedure and is relevant because the categorised outlines develop in a similar, but reverse, order to those of the grains shown later in Fig. 4.13.

The advent of the scanning electron microscope has made the examination of the surface textures of particles possible. An important example of this is in palaeoenvironmental studies. The surfaces of quartz-sand grains have been related to their environment and the findings used to determine past environments. The photographs in Fig. 3.8, illustrate the kind of detail involved, but it does not seem possible to obtain this as an edge silhouette that can be quantified.

For this research, the approach seems to be to use detailed descriptive terms such as those given in the Figure and others such as 'arc-shaped steps', 'scratches and grooves', 'parallel steps' and 'dish-shaped concavities' (Goudie, 1981).

If shape is believed to be the cause of a particular problem of powder behaviour, the use of sized and shape-sorted material may, at least, provide confirmation and may possibly provide more help than that. Even when shape measurement is to be carried out later, it may be worth while to use shape sorting to obtain an idea of how wide is the range of shapes concerned. As is well known, some shapes of particle are able to pass through a sieve more easily than others, so primitive shape sorting is possible in almost any powder laboratory.

If serious shape sorting is to be carried out, one of two methods will normally be used. The first uses the fact that chunky shaped particles move about a vibrating surface by hopping, whereas platey particles move about by shuffling. As a result

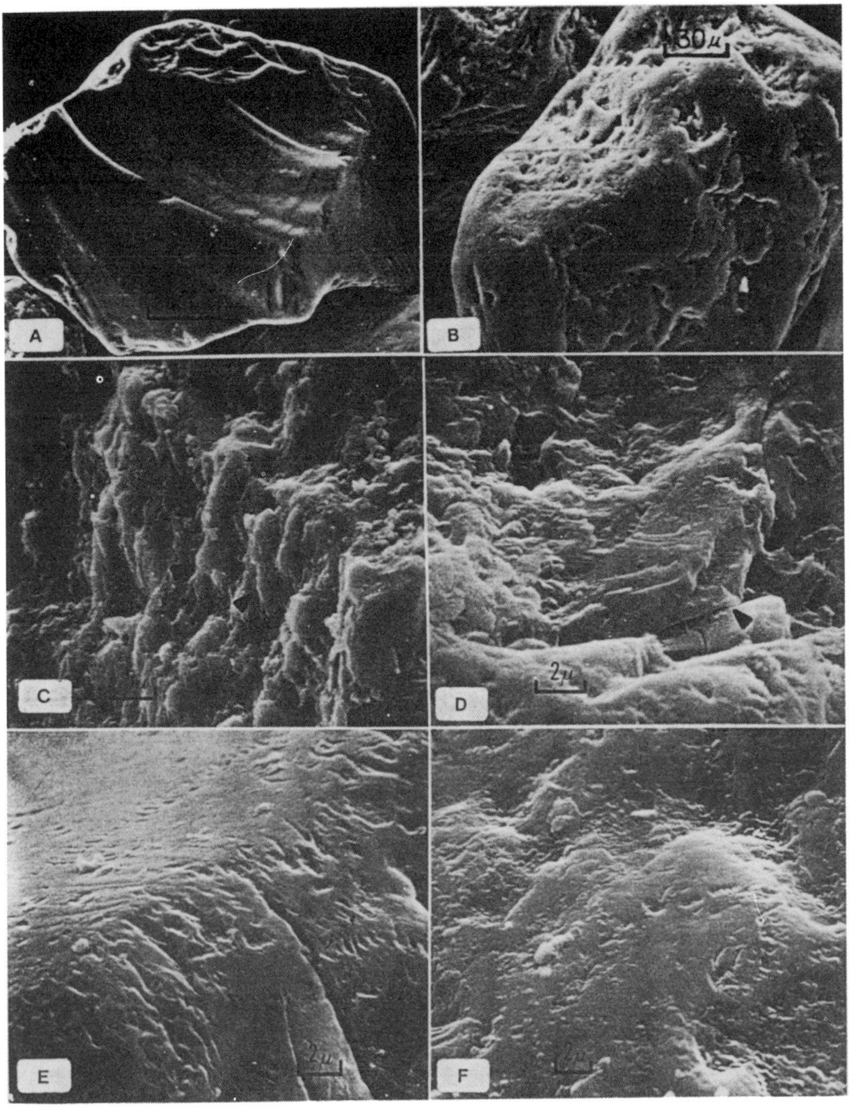

Fig. 3.8 Scanning electron micrographs of quartz grains illustrating textural features A = conchoidal fracture, B = blocky breakage, C and D = mechanically upturned plates, E = abrasion pits, F = mechanically formed V-shaped pits and crescentic percussion marks (Reproduced from H.W. Baker, *J.Sedim. Petrol.*, 1976, 46, pp 871-880, by permission of the Society for Sedimentary Geology)

of these two different modes of movement, a set of particles of different shapes will move on a suitably organised vibrating surface at different rates. This fact has been exploited by Ridgway and his colleagues in several investigations. Others, for example Riley (1968/69) and Viswanathan et al. (1984), have used travel across a rotating disc.

Ridgway & Rupp (1969) used a machine normally used for sorting industrial diamonds (Jeffrey-Galion Ltd., Johannesburg). There is a photograph of the machine in their second paper (1970/71), but the diagram in Fig. 3.9. illustrates the principle more clearly.

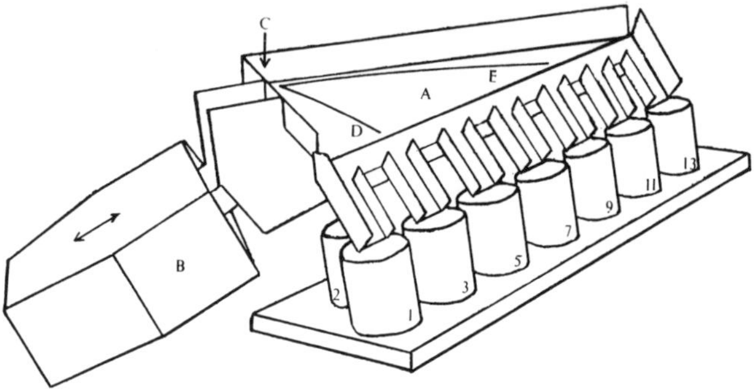

Fig. 3.9 A vibratory shape-sorting table. Particles fed on at C move across by gravity and ride up the deck by vibration generated by vibrator B

A sorting deck A slopes upwards in both the horizontal and the vertical direction. This provides the 'feed angle' and the 'sorting angle'. The deck is attached to a vibrator B operating in the direction shown. Particles deposited at C from, for example, a vibratory feeder follow short paths, such as D, if granular, and long paths, such as E, if flaky. Forward- and backward-facing chutes ensure that a particle will fall from the front edge into one of 13 collecting pots appropriate to its shape.

The authors write in their second paper:

'Trial and error adjustments of the feed and sorting angles and the vibration amplitude are necessary to obtain the best separation.'

A throughput of several kg day^{-1} is possible and material sorted in this way has been used for several investigations of the effect of particle shape on bulk properties. One publication was different in that it introduced a second method of shape sorting; this used slotted sieves produced by electroforming in nickel. Electroforming (Stork Veco, Eerbeck, The Netherlands) produces precise apertures of any desired shape down to a size measured in a few μm. Whiteman & Ridgway

(1988) report the use of quite large particles of the order of 500μm and conclude:

'Both the Jeffrey Galion deck and the slotted sieves were capable of sorting citric acid and sodium perborate according to shape. Each method has advantages and disadvantages which would influence the choice of method for a particular application. The deck sorted faster than the sieves for relatively large particles, and it can be operated continuously. By using decks with different surface properties, it is possible to separate a wide variety of materials by shape. Being almost infinitely adjustable, the deck is particularly good at shape-sorting particles with closely similar shapes.

'The main advantages of the slotted-sieving technique are that it sorts quantitatively with respect to two of a particle's three principal dimensions and it sorts more selectively than the deck. As the method requires no additional equipment apart from the one set of special sieves, it could easily be introduced into any laboratory.'

The Jeffrey-Galion table has also been reported in successful use to produce shape-sorted milled pasta and milled gelatine by Lenn & Holt (1982). These authors report the results of filming, at 1000 frames second^{-1}, the motion of particles on the vibrating deck and provide a physical explanation of the paths followed by different shapes. According to the authors, the shape-sorting table does not sort particles by shape but according to particle behaviour on a vibrating surface. From this behaviour a shape is inferred, for example granular particles are generally not carried with the vibration. The suggestion is made that, for the size range studied of +400-800μm, sorting was according to flakiness.

Meloy & Makino (1983) have used a stack of identical square-mesh sieves to shape-sort by residence time, and Whiteman & Ridgway provide a short review of some other shape-sorting methods that have been described in the literature. One paper that these latter authors omit is rather original and important (Nakajima et al., 1978). The research was intended to obtain particle-shape distributions without the need for measurements on individual particles. The particles were quite large, of the order of several millimetres, but the principles are generally valid provided that the powders can be sieved. The authors divided their material into size fractions with square-aperture sieves and then sieved the individual size fractions with specially fabricated sieves with long slotted apertures. As in the work by Whiteman & Ridgway, this sieving is in terms of the thickness and width of the particles.

It is not appropriate to reproduce the detailed mathematics here, but the authors are able to plot shape distributions using a 'shape factor' θ given by:

$$\theta = \tan^{-1} (\text{thickness} / \text{width})$$

By combining the techniques described in the two reports, it should be possible to produce a set of powders each with a restricted range of size and shape and with a known shape distribution. It will be realised, as the following Chapters are read, that such sets of powders could be valuable when testing the power and applicability of the shape-quantifying systems described.

4 Single-Number Classification

At one time, quantitative work on particles by microscopic examination involved dimensional estimation by micrometer eyepiece and area estimation by a grid micrometer eyepiece. This was replaced, for both size and shape studies, by measurements on an enlarged image, outline or silhouette of the powder particle concerned. It is assumed that the enlargement produces no distortion.

Before modern electronic aids became available, the enlarged shapes were obtained by one of the following methods depending, to some extent, on the size of the particles concerned :

camera lucida or drawing tube on a microscope,
projection microscope,
photomicrography,
slide projector.

In some cases, outlines were traced out on paper and the tracing was measured; in others, direct measurements were made on the projected images. A planimeter was often used for area measurement and an opisometer or map measurer for perimeter length. These techniques of measurement were, of course, commonplace in other disciplines, for example geography, and sometimes used for a similar purpose.

Today, relatively inexpensive aids are available to provide semi-automatic measurements on projected images. These generally utilise more information from a digitised outline of the planar form than previous methods; but, it will be seen later, not necessarily without some difficulties. It should also be noted that digitisation can produce different values for the perimeter and the area of a particle outline from those produced by an opisometer and a planimeter. This is because, as can be seen in Fig. 3.5, digitisation can provide only vertices for a polygonal representation of the outline.

Tsubaki & Jimbo (1979) used a scnic-digitiser with photographic prints of microscopic particles and Hawkins (1980) used a similar sonic-digitiser on rear-projected negatives of photomicrographs made with Polaroid instant film. A contemporary form of sonic-digitiser is shown in Fig. 4.1.

Fig. 4.1 A sonic-digitiser, showing the microphone array, touch control panel and stylus (Science Accessories Corporation/PMS Instruments Ltd., Maidenhead, Berks.)

A microphone array provides a transparent work area of more than 1200mm x 900mm, which may be mounted on many different display surfaces. The time taken for a sonic impulse from a stylus or a cursor is converted into digital coordinates. Graphic information is converted into digital form, so that lengths, areas etc. can be automatically obtained and recorded from the coordinates. This apparatus can be used with drawings and photographic prints as well as with projection systems, and the output format can be selected for compatibility with any computer.

A similar result may be obtained by an inductively coupled platen of the type shown in Fig. 4.2.

This particular unit cannot be used with rear projection, and is shown here set up for use with drawings or photographic prints, but it can be used with a microscope fitted with a camera lucida or drawing tube, or modified to operate equally well with video images. As with the sonic-digitiser, lengths, areas etc. are automatically computed and recorded, this time from coordinates obtained by the inductive coupling between a stylus and the platen. Other versions, which can be used with rear projection, are available with transparent platens.

The discrimination is good, but, depending on the magnification and the clarity

Fig. 4.2 An image analyser using a digitiser table (Kontron Electronik Ltd., Watford, Herts.)

of the image, is generally bettered in video, automatic-image-analysis machines, particularly those using a high-resolution television camera and associated equipment. However, it should be remembered that, with semiautomatic digitisers, the operator controls the input. The discrimination of an automatic video-analysis system may be seriously impeded by lack of contrast and by noise in micrographs and direct images. By combining human decision-making with computer technology, semiautomatic digitisers give measurements as reliable as the operator's judgment and can be used with most image media. This is a point made by the authors of a paper describing a semiautomatic image-analysis system devised primarily for biological work (Hayes & Fitzke, 1987). It is a low cost system using a small microcomputer and digitising pad (Taab Laboratories Equipment Ltd., Aldermaston).

It was reported recently (Lennard, 1990) that 'the availability of user-friendly scientific image analysis software for the Macintosh II has made the application of digital imaging techniques both practical and cost effective in many areas of research'. This is a valuable development in image analysis intermediate between manual digitisation and committed automatic-image-analysis machines. The author suggests that 'the addition of image processing capabilities to existing computers has provided a low-cost solution to improving data acquisition, analysis and the maintenance of a coherent image data base.'

Many of the single-number measures of shape described in this Chapter are available as a standard facility in automatic-image-analysis instruments and have been for several years. However, some of the recently introduced instruments provide the user with a wider choice. For example, with the Juliet 2 (The Shapespeare Corporation/Cottenham Instruments & Materials, Cottenham, Cambridge), it is possible to obtain some of the more detailed quantitative information described in Chapters 5 and 6.

With any system, photographs should be used with caution. In some cases it will be found that constant focusing adjustments are needed during the use of a microscope to allow for the varying heights of particles so that measurement is always made at the desired cross-section, and to distinguish overlapping particles.

Characterising an outline by a single number has two disadvantages: first, all single-number descriptors are ambiguous, more than one outline shape can have the same number, and secondly such a descriptor is imprecise, the outline shape cannot be reproduced from the number although, in a few cases, some visualisation of the outline is possible. Despite these demerits, single number systems will continue to be used. The allure is that they are both convenient to deal with mathematically and, in principle, can be targeted to the shape characteristic of interest. Williams (1965), when dealing with pebbles in three dimensions, wished to have 'only one parameter so that shape may be treated as a single statistical variable, studied in frequency distributions and correlated with other variables such as distance of transport and size of particle.'

Lloyd (1988) writes 'This paper reviews the many methods available for describing particle shape and develops the theme that the use of a single shape factor will never be sufficient and that emphasis should be placed on methods that use more than one.' This is an extreme view. It will be realised as this Chapter proceeds that different single-number descriptors can be useful as primary mathematical 'shape-sorters', used to correlate with step changes of bulk-powder behaviour, collected up to provide shape distributions and so on. In general, the measurements and computations necessary to generate single-number descriptors do not take long to complete even when performed manually. This should mean that large numbers of outlines are dealt with and thus the outlines are, at least, representative of any bulk powder concerned, even though the measurements are crude. The more complex the shape measurement, the fewer particle outlines seem to be measured, and thus the worth of the numbers representing the population in the powder bulk decreases.

The historical development of the measurement of particle outlines is an interesting topic to study, but unfortunately it does not provide an easily accessed text. It is expected that this book will be a starting-point for searches for appropriate outline-shape measurements. For this purpose, it is more convenient to divide the methods into groups. Such groups also facilitate the making of comparisons. The groups making up this Chapter are:

dimensional,
sphericity,
roundness,
perimeter.

It has not been possible to keep the groups entirely separate, and a further difficulty has been found with notation. Where quotation has made it essential, the original notation has been used, but, when possible, a uniform notation has been introduced.

Dimensional

The form of a large particle can be expressed by a parameter based on the longest, shortest and intermediate orthogonol axes. The review by Barrett (1980) can perhaps not be bettered for information on this aspect; there are adequate references, and the several parameters used since 1920 are detailed in tables. Another useful paper on triaxial measurements in practical use is that of Davies & Karuhn (1973). When these three values are available, volume may be invoked in numerical definitions of large-particle shape (Heywood, 1937). It should be noted here that Heywood states categorically that 'it is essential to distinguish between the shape and the proportions of a particle. Shape may be defined by the closeness with which a particle approximates to certain geometrical shapes such as ellipsoids, prisms or tetrahedrons. The proportions of a particle are quite distinct factors from the shape; thus either a flat or an equidimensional particle may have any of the geometrical shapes mentioned above.'

A critical and detailed review (Flemming, 1965) ends with a positive outcome : 'thirteen dimensional parameters are suggested as necessary to describe a particle, and a range of form indices is derived from these.' However, as it is generally accepted that, under the microscope, small particles must be dealt with in two dimensions, the topic might be expected to be left at that. The fact is that one reference in the table referred to above merits attention. Barrett concerns himself only with the fact that Aschenbrenner (1956) uses the orthotetrakaidekahedron to represent the shape of a rock particle rather than the ellipsoid introduced by Krumbein (1941). There is no mention of the fact that Aschenbrenner had introduced photogrammetry into the study of particles down to a size of 40μm; that is, he could measure the required three dimensions of particles that could not be handled directly. This was done by constructing a stereoscopic camera of elegant simplicity. This is illustrated in Fig. 4.3. taken from the technical paper by Aschenbrenner (1955).

Oblique illumination was used and a pair of stereoscopic photographs taken of each group of particles, preferably-size graded for satisfactory focusing. The particles were strewn on the sample plate surface, which had a scale grid thereon. Enlarged prints, or preferably, according to the author, transparencies, were aligned on a commercial mirror stereoscope and the appropriate dimensions measured.

This seems to be a unique and successful attack on the problem of small-particle shape, and it is surprising and disappointing that it has not been developed.

To return to two-dimensional outlines or silhouettes: from the measurement of the greatest length (L) and greatest width at right-angles (B) of a grain projection, a measure of the degree of anisometry, the 'elongation' of a grain, can be obtained (Schneiderhöhn, 1954). It is expressed as:

$$\text{elongation} = L/B \text{ or } B/L$$

This is particularly helpful with fibres. Wylie & Schweitzer (1982) say that it is important that careful attention be paid to the dimensional characteristics of mineral particles that occur in the workplace since there is evidence that the size and

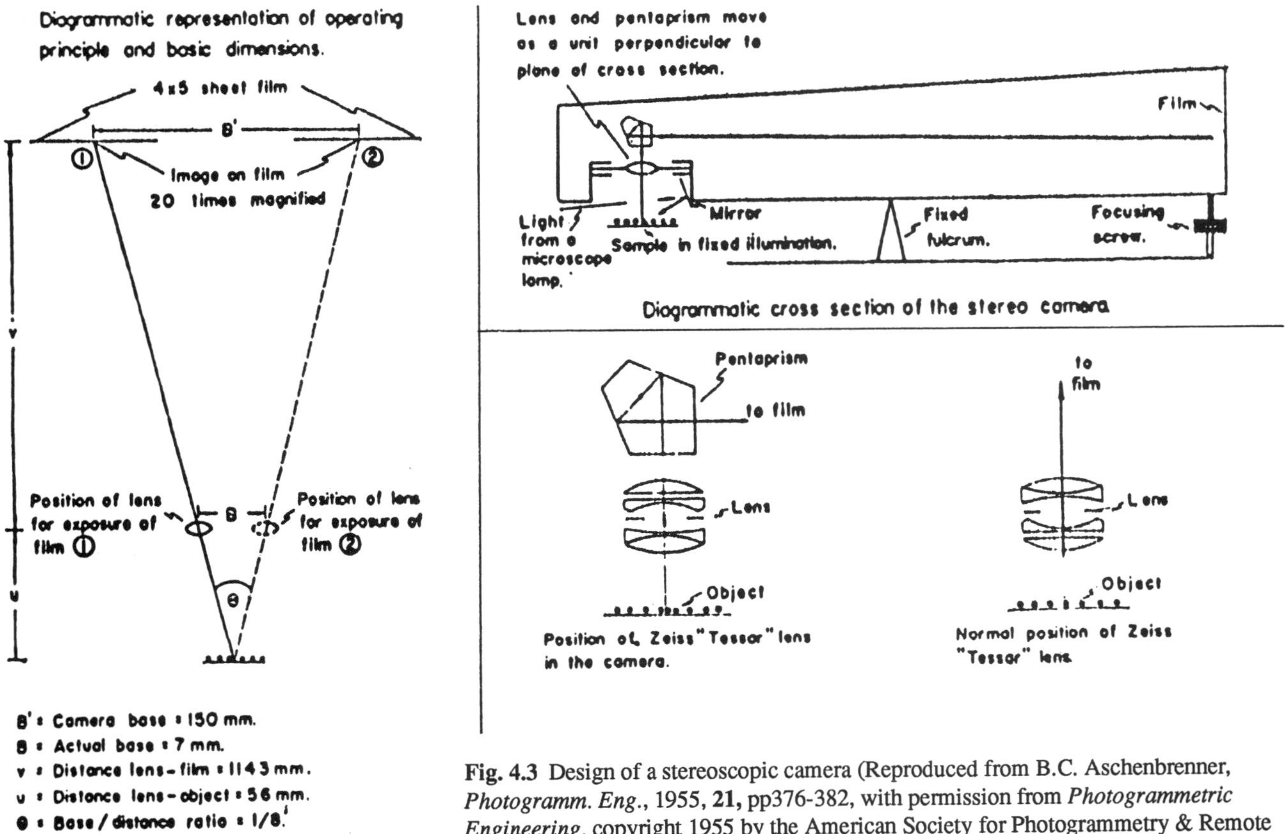

Fig. 4.3 Design of a stereoscopic camera (Reproduced from B.C. Aschenbrenner, *Photogramm. Eng.*, 1955, **21,** pp376-382, with permission from *Photogrammetric Engineering*, copyright 1955 by the American Society for Photogrammetry & Remote Sensing)

shape of such particles are major factors in determining their carcinogenic potential. The shape-characterisation method used was described by Siegrist & Wylie (1980) and used by them to distinguish asbestos from other mineral particles. The lengths L and widths B of a random sample of particles are plotted on log-log scales. The regression of log width on log length is of the form:

$$\log B = F \log L - b$$

The slope of the least-squares line, F, is taken as an index of fibrosity.

Fig. 4.4 is taken from the paper by Wylie & Schweitzer and shows how different the fibres can be, but it will be noted in Chapter 6 that such doubly logarithmic plots with least squares lines are not, necessarily, the best technique to use.

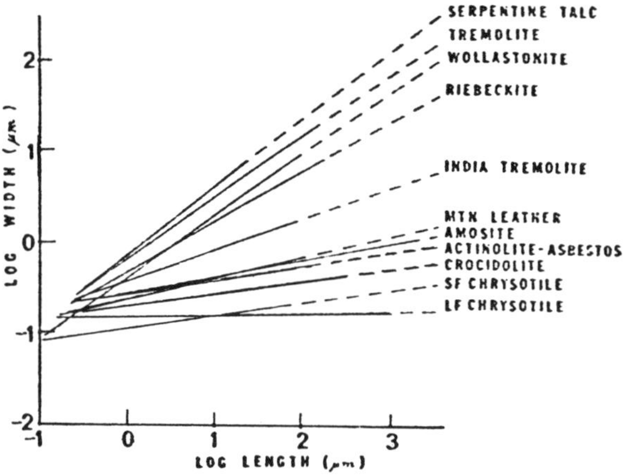

Fig. 4.4 Linear least-squares regression lines showing log width versus log length for fibres (Reproduced from A.G. Wylie & P.Schweitzer, *Environ. Res.*, 1982, **27**, pp.52-73, by permission of the authors and Academic Press Inc.)

Hagerman (1933) plotted the B/L ratio as ordinate and L as abscissa for grains on their greatest developed face. As plotting continued, the boundaries of the distribution field began to appear. In this particular work, four groups were concerned:

small, equiaxial grains,
small, oblong grains,
large, equiaxial grains,
large, oblong grains.

The kind of result to be expected is shown in Fig. 4.5, wherein the values plotted have been obtained from measurements made on the silhouettes given by Heywood (1945) in a paper on the measurement of particles and reproduced in Fig. 3.1 in the previous Chapter.

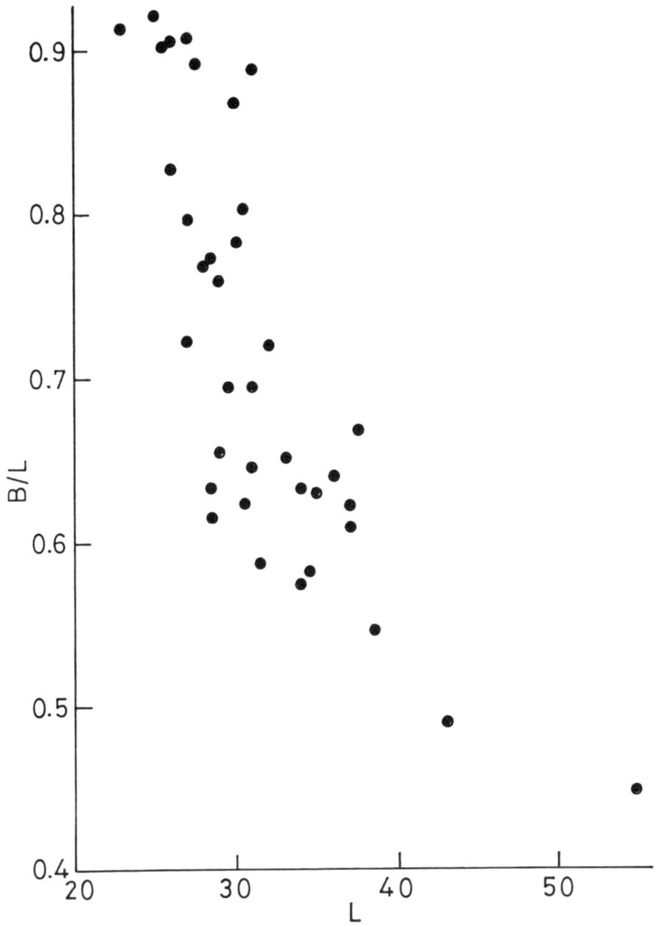

Fig. 4.5 Breadth/length plotted against length for the 38 particle silhouettes illustrated in Fig. 3.1

Where microscopic size is concerned, because particle outlines are not generally geometric figures, a set of conventional, so-called statistical, diameters have come into being; they require statistical averaging to obtain some quantity that can be related to a physical parameter of interest. The most popular ones seem to have been those that can be measured with a minimum of decision-taking. For the following two measures, it is assumed that the two-dimensional images are randomly aligned and the measures must be taken parallel to an arbitrarily fixed direction.

As shown in Fig. 4.6 Martin's diameter is the distance between opposite sides of the particle measured crosswise of the particle and on a line bisecting the projected area. Feret's diameter is the distance between two tangents on opposite sides of the particle. Kaye (1968/69) discusses the correct and incorrect use of these measures in some detail.

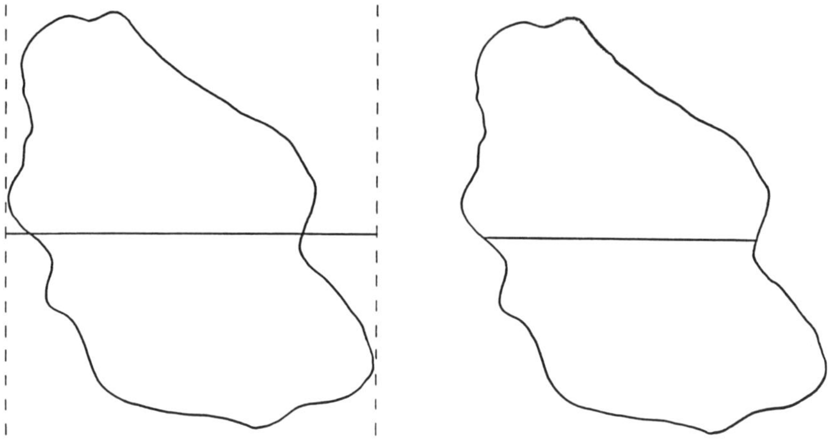

Fig. 4.6 Two statistical diameters

It will be seen in Fig. 4.6 that, for the same outline, the two diameters are different, and it will be realised that different outline shapes will have different effects on this. This was the subject of debate at one time. Some authors expressed dissatisfaction with the Feret diameter and recommended that its use should cease (Herdan, 1960). This was an error based on a specific piece of research by Heywood (1945).

The ratio (Martin's diameter) / (Feret's diameter) has been used to give a profile shape factor, but with no great success (Church, 1968/69). The same ratio was suggested as useful in an early article on features of an image-analysing computer (Gibbard et al., (1972).

In a paper reporting continuing investigations into practical techniques for the best use of a then relatively new commercial video-image-pattern-analysis system, McCarthy (1976) gave uses for the elongation ratio. This ratio was defined :

$$\frac{\text{Feret's diameter in whichever direction maximises it}}{\text{Feret's diameter in whichever direction minimises it}}$$

This choice was made because of its low sensitivity to image quality, in particular to signal noise.

Examples of its use include deviation from circularity of the outlines of glass spheres and of large elongation ratios for crystals viewed with a transmission electron microscope. The latter task introduced the concept of a mathematical 'shape-sorter'. In the paper, crystals appear as elongated rectangular shapes and agglomerates of crystals as irregular shapes. It is easy to arrange for a computer program to partition such measured values.

When employing semiautomatic, manual measuring, two different types of image are easily separated; but for automatic operation acceptable values of the elongation ratio can be used to instruct the machine to ignore the agglomerates, say, when obtaining, for example, size distributions.

The dimensions that McCarthy used for his elongation ratio were chosen for a different reason from those put forward by Heywood (1937). Heywood chose to take the breadth as the basic dimension, and it was defined as the minimum distance between two parallel lines tangential to the projected outline of the particle when placed in the most stable position. This definition was chosen because the above dimension is the deciding factor with regard to the passage or retention of the particle by a given aperture of a round or square-holed sieve mesh. The length of the particle is defined as the distance between two tangents to the projected outline of the particle drawn perpendicularly to the tangents defining the breadth. This is illustrated in Fig 4.7.

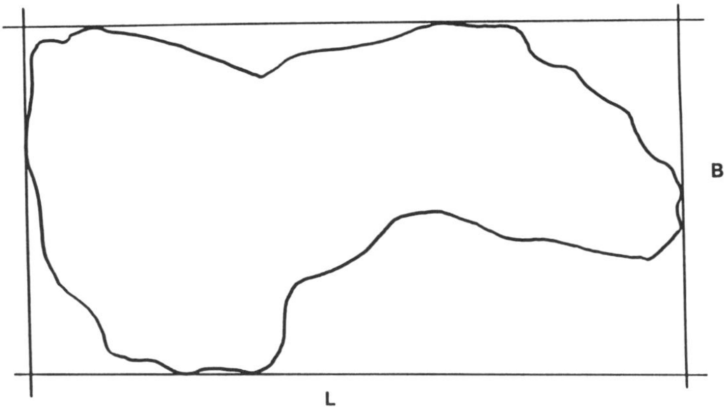

Fig. 4.7 A rectangle drawn around a particle outline to the Heywood (1937) specification

It will be seen that the projected outline of the particle is, by this method, thus circumscribed by a rectangle of breadth B and length L. The length ratio is L/B, and Heywood also used a projected area ratio defined as:

$$\frac{\text{(projected area)}}{\text{(area of rectangle } BL)}$$

The use of such a simple, dimensionless measure of outline shape as length/breadth would seem to be very restricted. This is not necessarily so. The length and breadth of leaves at different stages of growth, from a variety of plants, were measured, and it was found that for any one species the values fell closely on a straight line, $\log L = k \log B + b$, where k and b were constants (Hammond, 1941). If $k = 1$, the leaf grows without changing shape; if $k <$ or > 1, the enlargement of the leaf is accompanied by a change in shape. There are other papers on the morphogenesis of leaves treated in this manner; an interesting one is that by Ashby (1948).

In geomorphology, examples of such use may be found. One frequent example is the length-width ratio of drainage basins (Strahler, 1958). This is illustrated in Fig. 4.8, from a text by Krumbein & Graybill (1965).

It will be noted that there is an illustration of an alternative measure, called the drainage-basin circularity, which is another dimensionless value calculated from the ratio:

$$\frac{\text{(basin area)}}{\text{(area of circle of same perimeter)}}$$

A circular basin would have an index of 1, and other shapes would have an index from 0 to 1 (Miller, 1953). Similar measures are used in powder-particle studies.

Sphericity

It is unexpected to find the term 'sphericity' so often used when dealing with two-dimensional shapes, but it is so used and can cause confusion. Also confusing in some earlier literature is the incorrect use of the term 'roundness'. This generally occurs in papers written before that by Wadell (1932), who differentiated between shape and roundness. Roundness is a matter of the sharpness of corners, and shape has to do with the form of the particle. These are two independent variables. A cylinder with hemispherical ends is as perfectly rounded as a sphere, but has a different shape from a sphere; the cylinder has a different 'sphericity' in the terminology developed by Wadell.

In his original work on large particles, Wadell (1932) gave the term sphericity to the ratio:

$$\frac{\text{(surface area of a sphere of the same volume as the particle)}}{\text{(actual surface area of the particle)}}$$

Wentworth (1933) raised several objections to this measure. Like other geologists working on pebble shape at the time, Wentworth was interested in abrasion or wear. His solution to the problem of measuring 'roundness' was to use one of two ratios:

$$\frac{\text{(radius of curvature of the most convex part)}}{\tfrac{1}{2}\,\text{(longest diameter through the most convex part)}}$$

or

$$\frac{\text{(radius of curvature of the most convex part)}}{\text{(mean radius)}}$$

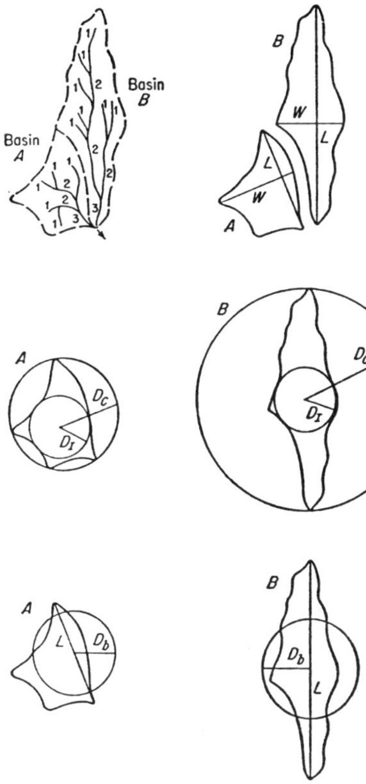

shape definition	basin A	basin B
Width/length, W/L	0.75	0.28
Diameters of inscribed and circumscribed circles, D_I/D_c	0.55	0.26
Diameter of circle with basin area vs. basin length, D_b/L	0.73	0.47
Circumference of circle with basin area vs. basin perimeter, C_b/P	0.83	0.62

Fig. 4.8 Ways of expressing drainage-basin shape (Reproduced from W.C Krumbein & F.A. Graybill, 'An Introduction to Statistical Methods in Geology'. McGraw-Hill Inc., N.Y., 1965, by permission of the publishers)

Wentworth (1919) was measuring quite large cobbles, but the measures are equally applicable to outlines. In his comments on 'sphericity' (for which he preferred the term 'compactness') ,Wentworth wrote:

'In view of the importance of the surface area ratio, long recognized, description of laboratory methods for making this and other related measurements with reasonable economy of time is awaited with interest.'

This, so called, 'true sphericity' was almost impossible to obtain even for large particles because of the difficulty of measuring the surface area of an irregular solid, and Wadell adopted a working formula that approximates to the true sphericity (Wadell,1934):

$$\frac{\text{(diameter of a sphere of the same volume as the particle)}}{\text{(diameter of circumscribing sphere)}}$$

It will be realised that the formula easily converts to one referring to the particle outline (Wadell, 1935):

$$\frac{\text{(diameter of a circle equal in area to that of the particle outline)}}{\text{(diameter of the smallest circle circumscribing the particle outline)}}$$

This is a formula very similar in principle to that put forward rather earlier by Tickell (1931):

$$\frac{\text{(area of the particle outline)}}{\text{(area of the smallest circumscribed circle)}}$$

There were two important differences: Tickell was using his ratio as the only shape measure, which he called 'roundness', and made his measurements on the outlines of particles presented with random aspects; Wadell was using two measures and always made his measurements with the grain resting on its larger face.

In a short summary of earlier work, Riley (1941) in a paper on 'projection sphericity' cites Pentland (1927), Cox (1927), Tickell (1931) and Wadell (1935). The formula given for Pentland is :

$$\frac{\text{(area of the particle outline)}}{\text{(area of circle with diameter equal to longest length of outline)}}$$

and that for Cox:

$$\frac{4\pi\ \text{(area of the particle outline)}}{\text{(perimeter of the particle outline)}^2}$$

However, the first three authors called their measures 'roundness' not 'sphericity'. After noting that these methods were 'so time consuming that very few workers have had the energy to undertake detailed studies', Riley proposed a new, quicker measure called the 'inscribed circle sphericity':

$$\frac{\text{(diameter of the largest inscribed circle)}}{\text{(diameter of the smallest circumscribing circle)}}$$

The inverse of this measure gives a value close to, although not generally equal to, the aspect ratio defined earlier.

The ambiguity that can arise with single-number measures is easily illustrated with inscribed-circle sphericity. In Fig. 4.9, all six outline shapes have the same value of approximately 0.33.

Fig. 4.9 Six outline shapes with the same value of inscribed-circle sphericity

A pragmatic approach to the estimation of Wadell sphericity was taken by Rittenhouse (1943), who produced standard reference charts for visual comparison. If these are to be used, a report on a careful and rigorous investigation of the sensitivity of the technique to differences between operators should be consulted

(Rosenfeld & Griffiths, 1953). However, the fitting of circles to particle outlines is not difficult, particularly if the apparatus due to von Endter & Gebauer (1965) is used to obtain the equivalent circle diameter. This is the more difficult of the two diameters and requires practice on particle outlines of known area.

It is important to note that the term 'sphericity' is not confined in the literature to the Wadell definition, and it is not clear why, for particle outlines or silhouettes, the term 'circularity' is not always used. Fudging the issue by expressions such as 'two-dimensional sphericity' is not satisfactory. Wadell, himself, used the expression 'degree of circularity', defined as:

$$\frac{\text{(perimeter of circle of same area as outline)}}{\text{(perimeter of outline)}}$$

The reciprocal of this expression is referred to by Horton (1932) as the factor 'compactness', given by the ratio :

$$\frac{\text{(perimeter of a drainage basin)}}{\text{(perimeter of circle of same area as drainage basin)}}$$

Roundness

So far as a silhouette or an outline of a particle is concerned, the idea of roundness is concerned with the sharpness or smoothness of the perimeter. This was illustrated in the previous Chapter, and Fischer (1933) used a straightforward method to quantify different types. A central point in the outline was taken, and angles around this point that were subtended by the straight or noncurved parts of the profile were measured. This is illustrated in Fig. 4.10.

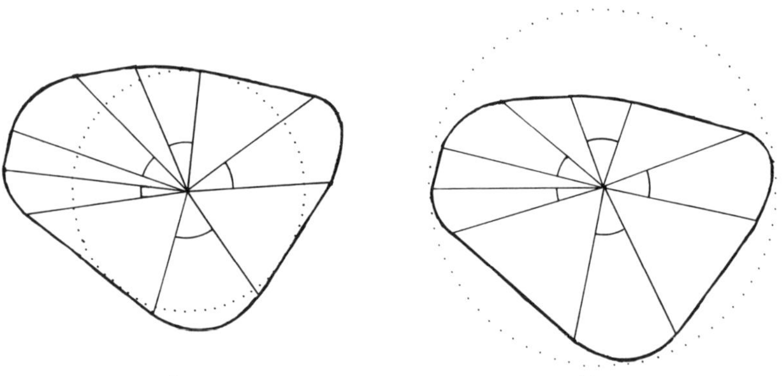

Fig. 4.10 Fischer's method of angularity computation.
A= inscribed circle; B= circumscribed circle.
Both centres give an angularity of approximately 0.42

To express the angularity value, Fischer used either the ratio:

$$\frac{\text{(sum of angles subtending noncurved parts of outline)}}{360°}$$

or the ratio:

$$\frac{\text{(sum of angles subtending convex parts of the outline)}}{\text{(sum of angles subtending plane parts of the outline)}}$$

The choice of the central point is left in the air by Fischer. The centre of either the inscribed or the circumscribed circle could be used, but, as will be seen in the next Chapter, modern equipment permits the use of the centroid of area, which is more satisfactory.

The Wadell concept of roundness tackles the same features of the particle outline as Fischer. The method is indicated in Fig. 4.11, wherein the radius of curvature (r) of each of the N projections on the perimeter is measured, together with the radius of the maximum inscribed circle (R).

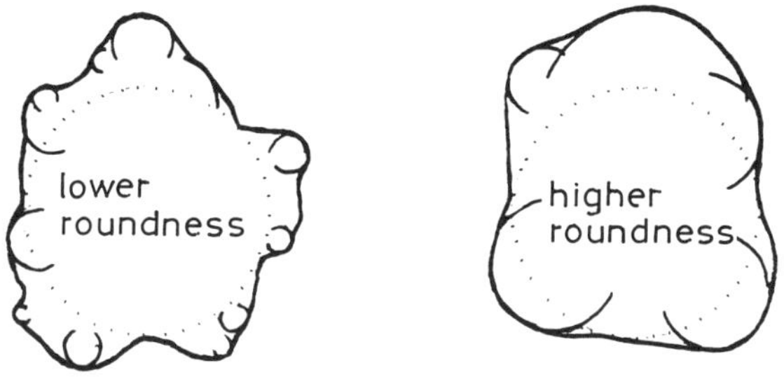

Fig. 4.11 Wadell's method of roundness measurement. The outlines have the same-diameter inscribed circle

Wadell considered several combinations of the data, but the usual method used (Wadell, 1935) is to divide the average of the radii r of the N corners of the grain image by the radius of the maximum inscribed circle R, that is:

$$roundness = \frac{\sum r}{N} \cdot \frac{1}{R}$$

It will be realised that the system gives a maximum value of unity and a zero value for angular projections. The radii for 'roundness' were measured using a transparent set of concentric circles, but the process was slow and the fitting of arcs to the particle outlines is also subjective. For these reasons, several workers devised a less demanding comparison system that was relatively rapid to use and permitted larger numbers of particles to be measured, and thus permitted more reliable sampling. Charts were made showing photographs or silhouettes of real or model particles of various degrees of roundness, together with the computed values of roundness by the Wadell system.

Russell & Taylor (1937), placed particles in five classes based on comparison with photographs of type grains. Pettijohn (1957) modified the values of the scales slightly to facilitate their use and also used silhouettes instead of photographs.

Powers (1953) who used photographs of clay models, as was illustrated in Fig. 3.5. introduced an extra descriptive term to provide more discrimination and also altered the class limits. The three schemes are given in Table 4.1.

Table 4.1 *Degrees of roundness: Wadell values.*

Grade terms	Russell & Taylor		Pettijohn		Powers	
	Class limits	Arithmetic midpoint	Class limits	Geometric midpoint	Class limits	Geometric midpoint
Very angular	N/A	N/A	N/A	N/A	0.12-0.17	0.14
Angular	0.00-0.15	0.075	0.00-0.15	0.125	0.17-0.25	0.21
Subangular	0.15-0.30	0.225	0.15-0.25	0.200	0.25-0.35	0.30
Subrounded	0.30-0.50	0.400	0.25-0.40	0.315	0.35-0.49	0.41
Rounded	0.50-0.70	0.600	0.40-0.60	0.500	0.49-0.70	0.59
Well rounded	0.70-1.00	0.850	0.60-1.00	0.800	0.70-1.00	0.84

It might be thought, from the wide use of the Wadell concept of 'roundness', that this was generally accepted. According to Piper (1971) this was not so. He cites Cailleux (1947), who used only the sharpest corner, as being followed by most French geologists, with Wadell as being followed by most American geologists; but points out that another measure of 'roundness' is the proportion of convex to concave parts as exemplified by the method due to Szadeczky-Kardoss (1933). According to Schneiderhöhn (1954), Wentworth used a ratio for outlines different from that he had developed for cobbles. This is illustrated in Fig. 4.12. in which it will be seen that the greatest length L of the outline is recorded, together with the greatest breadth B perpendicular to the length line. A firm value for a 'mean radius' is thus available. 'Roundness' was expressed by Wentworth as the ratio:

$$\frac{\text{(radius of curvature of the most convex part)}}{(L+B)/4}$$

The ratio used by Cailleux was similar:

$$\frac{\text{(radius of curvature of the most convex part)}}{L/2}$$

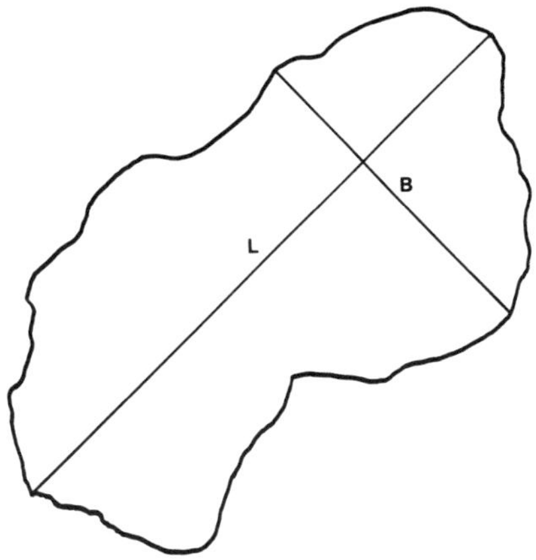

Fig. 4.12 The greatest length of an outline and the greatest breadth at right-angles to that length

For studying very angular material, Piper says Lees (1964a) used the angle between faces. Now, although 'angularity' might be considered a distinct concept rather than an absence of roundness, the paper by Lees contains a useful caution. The author points out and gives references to the particular attention devoted to the measurement and influence of the two parameters sphericity and roundness. 'While these parameters have generally been considered to define adequately the two main shape characteristics (overall shape and detailed surface contour) of most natural sedimentary particles of the pebble and sand sizes, they have been found to be respectively insufficient and unsuited to the shape analysis of particles produced by crushing. Such particles are often of sub-cubical or near parallelepiped shapes contrasting markedly with the more sub-spherical, ellipsoidal or pear-shaped forms of particles which have been affected only by attritional forces.'

Piper used an early form of digitiser and put forward the proposal that 'a two-dimensional measure of 'roundness' can be obtained from the distribution of angles between digitised segments of a grain'. Pickett et al. (1991) have used the advanced electronics now available to introduce a more powerful development in which the angles are measured with decreasing polygon line lengths. When the average angle is plotted against the line length, curves are produced that represent the unevenness of the outlines concerned.

A set of outlines from Schneiderhöhn (1954), who used the Russell & Taylor/Pettijohn classification, is given in Fig. 4.13. Schneiderhöhn made careful measure-

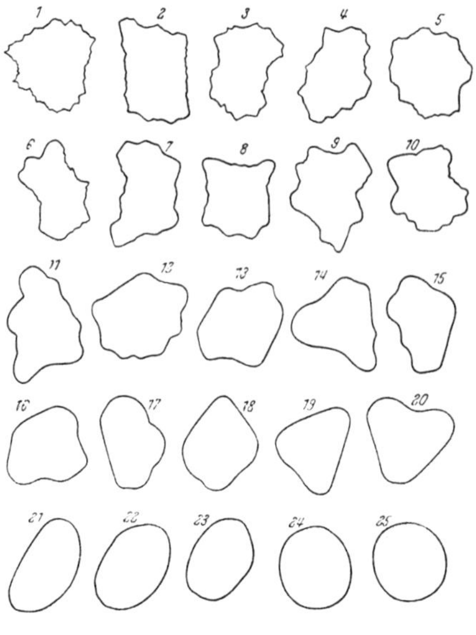

Fig. 4.13 Grain projections 1-25 used for measurements (Reproduced from P. Schneiderhöhn, *Heidelb. Beitr. Miner. Petrogr.*, 1954, **4,** pp.172-191, by permission of Springer-Verlag, Heidelberg)

ments of this representative set to compare the single-number methods then available. Some of the results are reproduced in Table 4.2. The author concludes from these values and those from other methods, including that of Szádeczky-Kardoss (1933), that the work reveals a need to find answers to various different difficulties occurring with all the methods used. The paper is an important one to read if it is proposed to use single number classification.

It was mentioned in the previous Section that Rosenfeld & Griffiths (1953) evaluated the limitations in the estimation of the sphericity of grains by visual-comparison charts. These authors did the same for roundness. They used seven students as operators assessing 21 quartz grains from a beach sand. The grains encompassed a variation of from 0.71 to 0.93 in sphericity and from 0.10 to 0.80 in roundness. Statistical analysis showed that it would be prudent to measure sphericity rather than estimate it, but that both sphericity and roundness could be estimated satisfactorily with care. The authors call attention to the fact that differences between operators were significant.

Powers, in his paper, criticises the use of silhouettes because 'the comparison characterizes a three dimensional particle by a two dimensional reference although the three dimensional aspect of small particles is easily observed under the microscope by raising and lowering the focal plane through the full thickness of the particle. A more complete picture of the roundness of a particle is obtained in this

Table 4.2 *Measurements of roundness and form.*
Recast from P. Schneiderhöhn, *Heidelb. Beitr. Miner. Petrogr.*, 1954, **4** pp.172-194, by permission of Springer-Verlag, Heidelberg, Germany

Grain number	Russel & Taylor Pettijohn classification	Wadell $\dfrac{\left(\frac{\Sigma r}{R}\right)10^2}{N}$	Wentworth $\dfrac{r_1.10^2}{R}$	Cailleux $\dfrac{2r_1.10^2}{L}$	$\dfrac{\text{Length}}{\text{Breadth}}$	Wadell $\dfrac{d.10^2}{D}$
1		0.00	0	0	1.07	81
2		0.02	0	0	1.50	74
3	Angular	0.2	0	0	1.43	80
4		0.4	0	0	1.63	78
5		1	0	0	1.04	86
6		15	5	3	1.61	73
7		16	2	1	1.70	75
8	Subangular	18	4	3	1.07	81
9		19	4	3	1.41	75
10		20	9	7	1.16	81
11		25	10	7	1.58	73
12		27	8	6	1.29	82
13	Subrounded	29	16	13	1.33	82
14		36	18	15	1.15	80
15		37	15	12	1.52	77
16		40	19	16	1.22	86
17		44	18	14	1.58	79
18	Rounded	46	27	22	1.30	82
19		51	33	27	1.29	80
20		59	42	37	1.11	83
21		63	45	32	1.92	72
22		68	55	46	1.48	83
23	Well rounded	78	56	43	1.45	77
24		89	82	78	1.11	95
25		95	90	86	1.08	96

manner.' Two different appraisal systems would be in operation with a microscope used in this manner, and that is unwise; the employment of silhouettes of particles presented in a random orientation is more self-consistent. Similar consistency is achieved if the size of the particles allows examination under the microscope by the method of rotating particles on a wire or thread 'microlathe', as described by Davies (1975).

Davies, in a detailed account of the background literature to a proposed representation of particle shape, calls attention to the fact that, in the early work on roundness, only nonconvoluted particle profiles were measured. This is true, but omits the more important point that there is a degree of opinion about where the texture of the profile begins. Mackey (1965) states 'Wadell divides the shape of the particle into two components, namely, sphericity and roundness. The former term is self explanatory, the latter is concerned with the surface texture of the particle.'

Piper (1971) considers the word 'roundness' an unhappy choice to express the concept put forward at the beginning of this Section and prefers 'roughness'. This author calls attention to the need to define the scale on which 'roundness' or 'roughness' is to be recognised. A year later, Whalley (1972) wrote:

'Texture is difficult to define and quantify. In part this is a scale problem because surface features on a larger scale become angularity and roundness. However, in practice, this is less difficult than it might seem in that there will usually be at least an order of magnitude difference between the two. Likewise, surface texture cannot be recognised on the projected outline of a particle and thus cannot be digitized.'

Barrett (1980) says that texture needs to be considered as a ratio of the type (texture size)/(particle size), which seems correct.

A measure called the 'roughness factor' or 'rugosity coefficient" has been used to express any lack of smoothness in a particle outline and involves the concept of a 'convex hull'. A convex hull is a minimum enveloping boundary fitted to a particle outline that is nowhere concave; two examples are shown in Fig. 4.14.

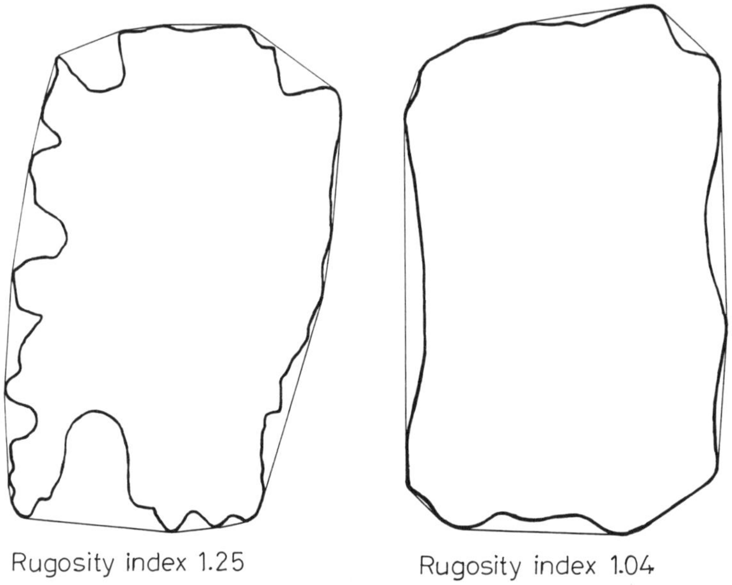

Fig. 4.14 Two particle outlines illustrating the convex hulls and the respective rugosity indices

Incorrect illustrations of convex hulls can be found in the literature; an elastic band should be imagined extended around the particle outline.

The rugosity coefficient is defined as :

$$\frac{\text{(perimeter of the particle outline)}}{\text{(perimeter of the convex hull)}}$$

The word 'rugosity' has been used before when air permeability and sieving on small quantities of powder have been used to measure average roughness and angularity (Robertson & Emödi, 1943). Rugosity was defined by these authors as the reciprocal of sphericity:

$$\frac{\text{(measured specific suface)}}{\text{(hypothetical surface area grains would have if they were spherical)}}$$

'Rugosity' has a specialised meaning of 'wrinkled', which is not the same as 'roughness', but would seem to be acceptable. Note, however, that, if rugosity is defined as the reciprocal of sphericity, it is not a measure of 'roundness' in the Wadell sense.

It will be realised that the rugosity coefficient is another example of a useful mathematical shape-sorter in automatic image analysis, but its use for other purposes should be after due consideration because there is a difficulty in terms of magnification and measurement discrimination. It is well known that the distance to be travelled between two towns measured as accurately as possible on a small-scale map will work out far smaller than the distance between the same two towns measured along the same roads on a large-scale map. This is, of course, because the twists and turns of the road are smoothed out on the small-scale map; the detail is missing. This was well put in a different context by Adamson (1960) when discussing the geometrical problem of the area of a solid:

'The difficulty in stating just what is meant by the 'surface area' of a solid, in one aspect, can be illustrated by considering the somewhat analogous question of what is meant by the 'length' of a particular section of coastline. Superficially, it seems easy to say that the length is simply the distance along the shore, between two points. Geography books, however, are prone to make such statements as 'because of its many indentations, the coast of Maine is actually some 2000 miles long'. It seems reasonable that the coast of Maine should be considered as being greater than that of a straight line drawn from border to border, and that the added length due to harbors, bays, points, etc. should be included. However, once one begins to consider irregularities, the problem is when to stop.'

Adamson goes on to illustrate the problem with Fig. 4.15. showing successive magnifications of a hypothetical stretch of coastline. With each successive magnification, the computed coastline length is increased. 'There seems to be no logical stopping point to the above sequence of successive enlargements if one is after some absolute value for the length of the shoreline.' This fact has been known to certain members of the population at large ever since maps at different scales were readily available, particularly to touring cyclists and walkers. However, the significance in certain areas of science seems to have been missed.

The problem is present with any expression involving the measurement of the length of a perimeter and needs to be considered in those factors in the next group that involve specifically measurements of length.The phenomenon has important mathematical consequences, of possible value in outline-texture measurements, which are discussed in Chapter 6.

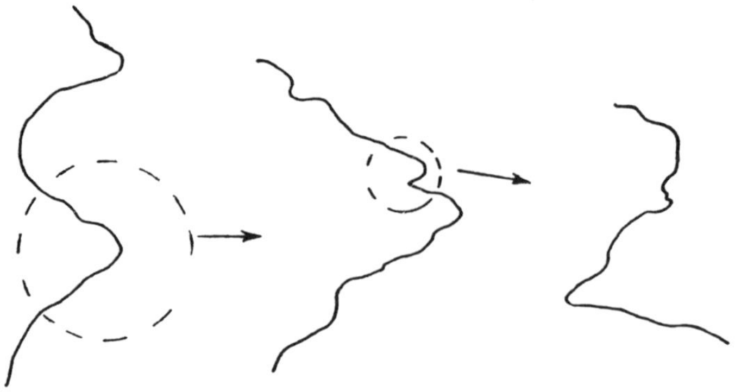

Fig. 4.15 Successive enlargements of a hypothetical coastline (Reproduced from A.W. Adamson, 'Physical Chemistry of Surfaces'. Copyright © 1960, John Wiley and Sons Inc., by permission of the publishers)

Perimeter

It was noted earlier that Cox (1927) put forward a measure for the 'circularity' of an outline that involved the perimeter. The reciprocal of this measure has become very popular in powder technology partly because of the facility with which it can be measured with automatic or semiautomatic image analysers. The measure today is generally called the circularity shape factor, given by:

$$\frac{(\text{perimeter of the particle outline})^2}{4\pi\,(\text{cross-sectional or projection area of particle outline})}$$

but Butler et al. (1990) use the expression 'morphic aspect ratio'.

Cox claimed that his measure was the same for all figures of the same shape, regardless of size; this is true only if a change of size does not alter the detail taken in when the perimeter is measured. The outlines in Fig. 4.16 are from the paper by Cox.

Unlike 'circularity' which is unity for a circle and less than unity for any other shape, the circularity shape factor is greater than unity for any outline shape other than a circle. There is, however, the inevitable ambiguity with the circularity shape factor. This is illustrated in Fig. 4.17. It wili be seen that the same values can come from differently shaped outlines. In many cases, this will not be a problem: the particles being measured will have a family likeness. When family likeness exists, not only will there not be a danger of ambiguity, but the circularity-shape-factor values will permit visualisation. This is illustrated in Fig. 4.18. by a set of ellipses. The Figure also includes a rogue shape, the value for which stands out.

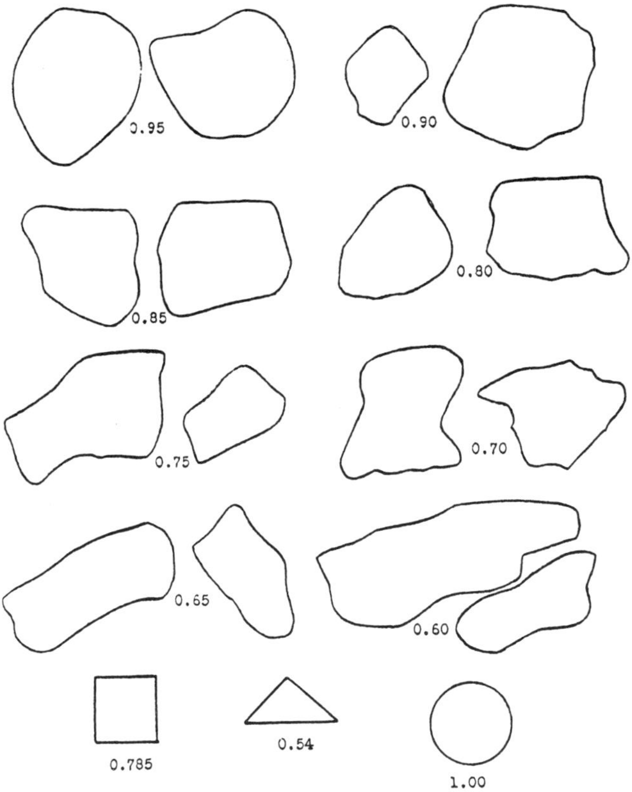

Fig. 4.16 Two outlines from each of a number of degrees of 'roundness', together with some geometrical forms for comparison (Reproduced from E.P. Cox, *J. Paleont.*, 1927, **1**, pp.179-183, by permission of the Paleontological Society)

This formula is called either 'circularity' or, very often, 'compactness' by researchers in 'computer vision' and is sometimes used in the form $(1\text{-}4\pi A/P^2)$, which equals zero for circles and approaches unity for the least-circular shapes. The literature of research into shape matching and shape classification for robot design or character recognition, for example, is extensive, and a helpful text to use is that by Levine (1985). This text is both detailed and well referenced. The ambiguity referred to above is illustrated by 20 different outlines, all with 'the same area and perimeter and hence the same circularity A_S.' A_S is given by $(1\text{-}4\pi A/P^2)$ as above.

In a study of the shape of national borders, Richardson (1961) used the formula:

$2\ (\pi \times \text{area})^{1/2} /\ (\text{perimeter})$

to judge the compactness of closed figures. This author drew 12 outlines of various

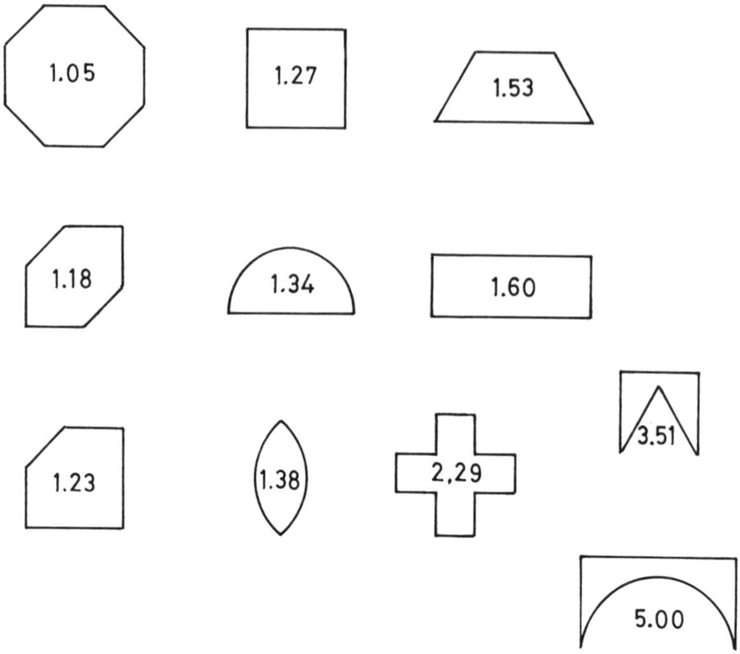

Fig. 4.17 Circularity shape factors for various outlines

shapes and had them ranked for compactness by 50 individuals, but did not publish any comparison with calculated values. It is not clear why Richardson did not choose the squared form of the formula. Both expressions are independent of the size of similar figures, and the values are unity for a circle; but the sensitivity is less in the Richardson version. For regular pentagons the approximate values are as follows, with the Richardson numbers given first:

triangle: 0.78, 0.61 ; square: 0.89, 0.79
pentagon: 0.93, 0.86 ; hexagon: 0.95, 0.90

Reverting back to powder literature, this method of dealing with the shape of a particle outline was put forward as one of three useful ratios by Hausner (1966). In this contribution to a particle-size-analysis conference, Hausner used the expression:

$$\text{surface factor} = (\text{perimeter of shape})^2 / 4\pi(\text{area of shape})$$

without acknowledging previous usage of, for example, the reciprocal by Cox. The other two ratios put forward by Hausner in his contribution involved drawing a

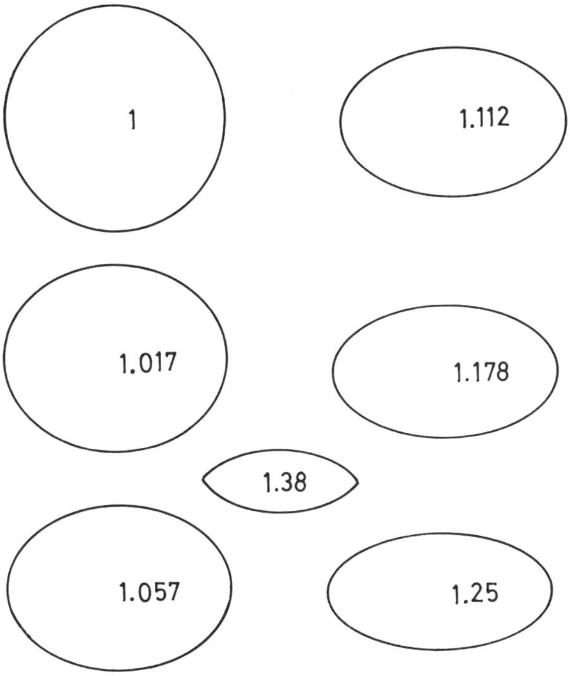

Fig. 4.18 Circularity shape factors for ellipses

rectangle of minimum area around the projection of a particle. Such a scheme is shown in Fig. 4.19, the rectangle of minimum area being found by trial and error.

According to Hausner, the extent to which the outline is elongated is given by a/b and the 'bulkiness' of the particle by (projected area of outline)/(area of rectangle ab). Each of these three ratios is ambiguous, but the three together do provide some visualisation. At the conference, Heywood pointed out that papers in earlier years by other authors had described similar procedures for defining particle shape; he also provided a short bibliography. However, as was described earlier, Heywood (1937) did not use a rectangle of minimum area, nor did he use his 'projected area ratio' as a bulkiness parameter.

Lloyd (1988) states that Medalia (1970/71) proposed the geometric factor widely used in image analysis for 'circularity' or 'roundness', where the circularity factor is given by:

$$\frac{(\text{perimeter})^2}{4\pi(\text{projected area})}$$

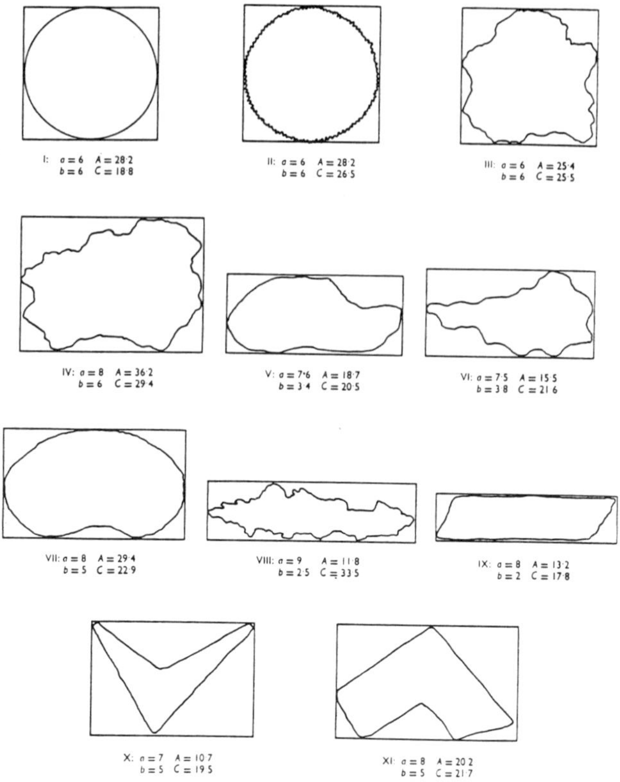

Fig. 4.19 A variety of particle outlines, showing the side lengths a and b of the rectangle, the projected particle area A and the circumference of this area C (Reproduced from H. H. Hausner, *Planseeber. Pulvermet.*, 1966, **14,** pp.75-84, by permission of Metallwerk Plansee GmbH, Reutte, Austria)

This, however, is the surface factor put forward by Hausner (1966, 1967)); and, in fact, the paper by Medalia introduces a new concept, which he calls the 'dynamic shape factor'. Medalia replaces a plane figure of arbitrary shape (i.e. the silhouette of a particle) by a radius-equivalent ellipse in the plane of the figure. This is illustrated in Fig. 4.20. taken from the original paper. The orientation of the ellipse is fixed by taking its geometric axes as (or parallel to) the principal axes of the particle. The length of each axis of the ellipse is taken as four times the radius of gyration of the figure about the other axis; thus, the radius of gyration of the ellipse about each axis is the same as that of the actual plane figure. The ratio of the longer axis to the shorter is taken as the anisometry of the figure. The ratio of the area of the ellipse to the area of the figure is taken as the bulkiness of the figure. This extension of the enveloping rectangle proposed by Hausner was only made possible

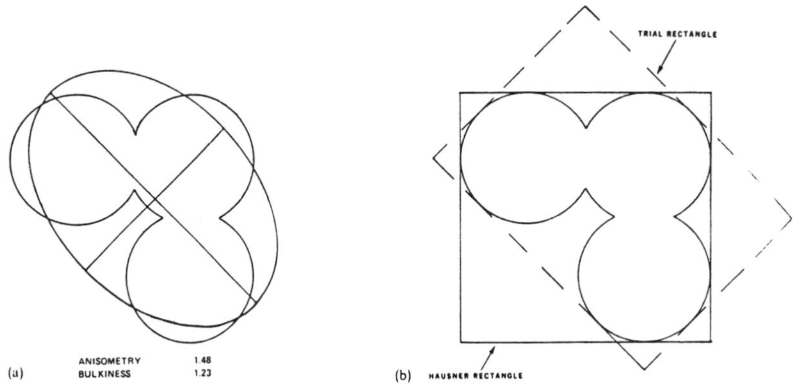

Fig. 4.20 Three-particle floc (a) with ellipse, (b) with Hausner rectangles (Reproduced from A. I. Medalia, *Powder Technol.*, 1970/71, **4,** pp.117-138, by permission of the author and Elsevier Sequoia S.A.)

when digitisers became available, since the coordinates of 'every' point on the outline must be determined and entered into a computer for the necessary calculations to be made.

Medalia gives his values for anisometry and bulkiness of a C-shaped figure and of seven reference shapes generated by the IIT Research Institute. These are reproduced in Fig. 4.21.

The facts that the coordinates of many points around the perimeter of a particle silhouette or outline may be obtained rapidly with modern equipment and that the values may be manipulated by computer have led to much activity to improve the quality and usefulness of particle outline characterisation. The enormous amount of numerical data that can be quickly collected has led to much work and many publications on the subject of data condensation.

A promising system that has been used in practice on fine material from a tumbling-ball mill uses two single number classifiers (Austin et al., 1990). Tsubaki & Jimbo (1979), mentioned at the beginning of this Chapter, used an index 'elongation', defined differently from that reported earlier (Schneiderhöhn, 1954). Their new definition is expressed as:

$$\text{elongation} = \frac{\text{smallest Feret diameter}}{\text{Feret diameter perpendicular to the smallest Feret diameter}}$$

This index is used by Austin et al., together with an original one. The fine particles are examined in a scanning electron microscope, which digitises any chosen outline into 1000 points. Every 50th point is taken to provide a 20-sided polygon. The authors develop the mathematics, but, in short, the mean contained angle for a polygon of n equal sides fitted into a circle is ϕ_c given by:

$$\phi_c = \pi(n\text{-}2)/n$$

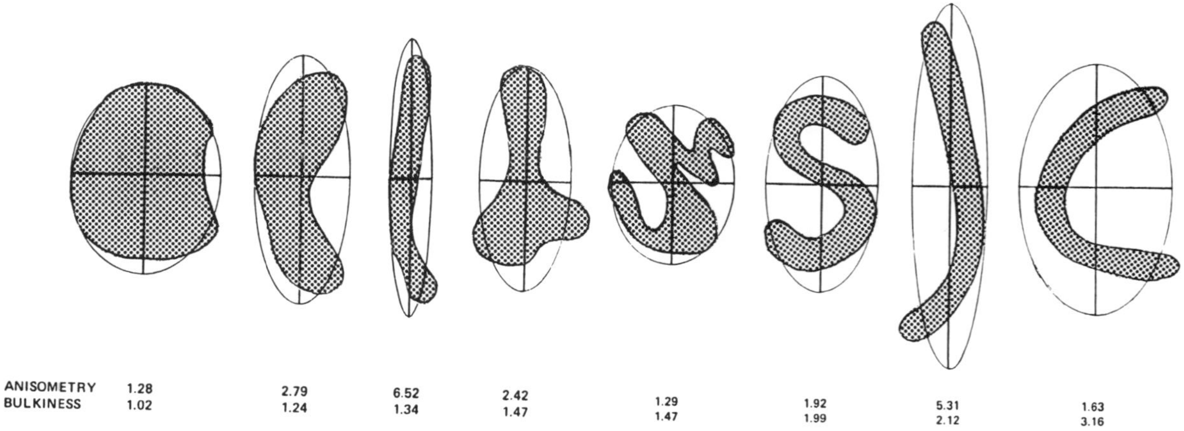

Fig. 4.21 Shapes illustrating anisometry and bulkiness (Reproduced from A. I. Medalia, *Powder Technol.*, 1970/71, **4,** pp.117-138, by permission of the author and Elsevier Sequoia S.A.)

The authors use an index 'angular variability' A, given by:

$$A = 1 + \sum_{k=1}^{n} (\phi_c/\phi_k - 1)^2$$

This is unity for a circle and increases as the particle outline becomes more irregular. The two indices are used together to partition the outlines, using 'elongation' horizontally and 'angular variability' vertically.

An informative paper on how digitised data can be usefully manipulated through the use of a Fortran program is that by Starkey & Simigian (1987). These authors use an X-Y recording microdensitometer to generate the data from a photomicrograph and 'compute the following parameters for each particle: area, length of the perimeter, long dimension and width at right angles to it, aspect ratio, centre of gravity, and orientation of the long dimension.' The program 'also fits regular shapes, rectangles, and ellipses, to the irregular particle boundaries.'

It is important not to be carried away by the amount of data that can be collected today. Beddow (1983) states, for example, that 'if we define the information content of a parameter on the basis of the number of x, y points of the profile that were used to define that parameter, we may develop the information of the sort given in the table.' (Table 4.3).

Table 4.3 *Information content of shape descriptions.*
(Reproduced from J.K. Beddow, *Particulate Sci. & Technol.*, 1983, **1**, pp.101-123, by permission of Hemisphere Publishing Corp., Bristol, Pennsylvania)

	Information content (number of x,y points)
Rugosity	All
D_m/D_f	4
Flatness ratio $M = (B/T)$	4
Elongation ratio $N = (L/B)$	4
Sphericity	All
Roundnes	All
Anisometry	Few
Bulkiness	Few
(Perimeter)^2/2 (area)	All
Morphic descriptions	All

Beddow continues 'Rugosity uses all of the x,y points whereas the ratio of Martin's diameter over Feret's diameter uses only four. The flatness and elongation ratios likewise use only four. Isometry and bulkiness use few x, y points and therefore they have a low information content. On the other hand, sphericity, roundness and perimeter squared over twice the area contain all of the x, y points and therefore have a higher information content.' This approach can be misleading, as a simple example will show. Consider the two particle outlines in Fig. 4.22. The two out-

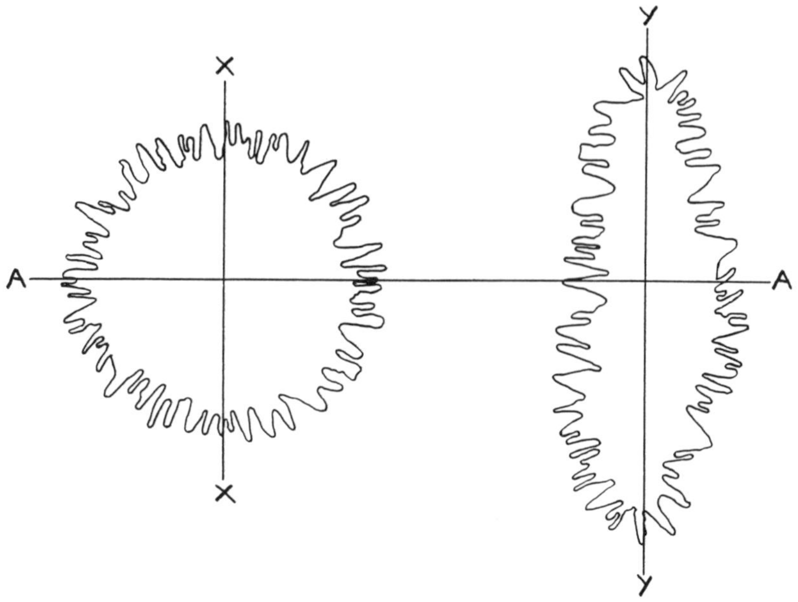

Fig. 4.22 Two particle outlines with similar rugosity indices but different aspect ratios

lines have similar values for the rugosity index; but if they be rotated through 180° about the axis *AA*, two solids are produced with a different behaviour when moving in a fluid for example. If the two outlines be rotated about the axes *XX* and *YY*, respectively, two solids are again produced with a different behaviour, but one will have a behaviour similar to that produced by the rotation about *AA*. In this example, more useful, i.e. relevant, information would probably have been obtained from elongation rather than rugosity.

Before moving on from single-number classifications, it must be pointed out that it is not clear from the literature how simple dimensionless ratios, for example of two lengths or two areas, came to be called 'shape factors' when they might be expected from mathematical usage to be called 'shape indices'; they are so called in some papers, but, in at least one paper (Riley, 1977), it is implied that the two terms are synonymous.

According to Herdan (1960), 'shape factors have a threefold function:

(1) They are the factors of proportionality between particle size determination results by different methods;
(2) They are conversion factors for expressing determination results in terms of an 'equivalent' sphere;
(3) They transform the second and third powers of the variable, particle diameter, into the particle surface and particle volume, respectively.'

A similar opinion is put forward in precise detail by Rumpf (1990).

Interesting accounts of the practical use of authentic shape factors are given by De Silva & Creasy (1983) and by Scott (1984). Another example of what would seem to be the correct use of 'shape factor' is in an early, but still helpful, paper on particle-size analysis (Fairs, 1943). The author writes:

'....all methods of size analysis involving the measurement of free falling speeds express the particle diameter in terms of the 'equivalent sphere'. The microscope, however, measures a projected area, which, in the case of non-spherical particles, must always be related to the two major axes, and it has been found, when using an ordinary linear scale graticule for size estimations, that the particle diameters obtained thereby are in general greater than those calculated from the corresponding free falling speeds. This necessitates the use of a 'shape factor' to correlate them with the Stokes diameter.'

When reading the 'shape' literature of many disciplines, remarks of the following type will be found: 'Undoubtedly the classic work on shape and form in any field is Sir D'Arcy Wentworth Thompson's 'On Growth and Form''. Despite a new, condensed edition of the book in 1942 and an abridged version in 1961, there are few references to it in the literature of powder science and technology (Hawkins, 1980). An important section points the way to dealing with the data problem arising from the use of polar coordinates to describe a two-dimensional outline. Sebestyn (1959) suggested characterisation of silhouettes by polar coordinates of their peripheries with the centre of gravity of the figure as origin. Separately, a similar scheme was used by Boyce & Clark for a study in urban geography (1964). However, these authors wished to use the data to rank the two-dimensional shapes of towns and cities and to compare them with specific geometric forms. The manner in which this was achieved does not seem to have appeared in powder literature and is as follows.

A point within the outline was taken as origin. It is sensible to use the centre of gravity of the shape, and this may be determined mechanically or mathematically. The distance from this centre to the outside edges of the shape was measured along equally spaced radials in a similar manner to that illustrated in Fig. 3.5 in the previous Chapter. The percentage of each radial distance with respect to the sum of all the radials was computed and each percentage subtracted from the percentage that each radial would be expected to have as based on a circle. Boyce & Clark state that the signs should be ignored and the numbers summed to give the shape index. Thus the index is expressed:

$$\sum_{1}^{n} \left| \left(\frac{r_i}{\sum_{1}^{n} r_i} \cdot 100 - \frac{100}{n} \right) \right|$$

For Milwaukee, the authors obtained an index value of 25 and for Dallas a value of 13. If the index numbers are determined for different shapes, a particular index value can be visualised. Milwaukee was similar to a rectangle (a value of 28) and not, like Dallas, a square (a value of 12); but, of course, the index does not permit reproduction of the shape. Boyce & Clark use the expression 'shape curve' for linear plots of radial number along the horizontal axis from 1 to *n*, and the percentage distance for each radial along the vertical axis. Some authors, when they plot radial distances against radial number, refer to the linear plot as the 'particle waveform'.

Boyce & Clark do not refer to the potential of their system for archiving and, although there is a reference to Sir D'Arcy Thompson, they may not have appreciated the significance of the following quotation:

'...on the outline of the simple leaf, whether ovate, lanceolate or cordate, we may superpose secondary sine-curves, of lesser period and varying amplitude after the fashion of a Fourier series...'

One year after the paper by Boyce & Clark was published, there was an account of the use of truncated Fourier series to quantify the outline of the human face (Lu, 1965), and this was followed independently by other publications illustrating the use of the procedure for different outlines. It has been suggested that this method of outline-shape quantification (which should be referred to as harmonic analysis) was, perhaps, overzealously promoted by some practitioners, but it is powerful and its virtues and vices are dealt with in some detail in the next Chapter.

5 Representation by Series

The suggestion by D'Arcy Thompson, quoted at the end of the last Chapter, that the shapes of leaves might be illustrated by sine curves after the fashion of a Fourier series does not appear to have had any impact whatsoever. However, the concept did eventually appear in several different disciplines more or less at the same time but, for some years, with no reference to one another.

Fox & Brazee (1971) make the important point that biological material, in the broadest sense, will have some recognisable form:

'If a biological structure were completely random in form, it would probably be impossible to numerically classify or even visually recognise the associated biological entity. The fact that humans can identify biological forms seems to be a good physical basis for expecting that meaningful mathematical representations of biological forms do exist.'

Chernoff (1973) used this fact when he created cartoon faces to represent points in k – dimensional space:

'The use of the face representation provides a promising approach for a first look at multivariate data which is effective in revealing rather complex relations not always visible from simple correlations based on two-dimensional linear theories. It can be used to aid in cluster analysis, discrimination analysis, and to detect substantial changes in time series.'

Six different measurements were made on 87 fossils, and cartoon faces were produced by computer using different eye, mouth, nose and face outline shapes. Three clusters of similar visual impact were apparent in the published illustration.

Lu (1965) was interested in quantifying the shape and symmetry of the human face. He states:

'By appropriate choice of coordinates and plotting technique, both the frontal view and the profile may be considered as single-valued curves defined over closed intervals; each consists of a set of systematically unequal magnitudes with a finite number of extrema and inflexion points'.

He continues:

'By choosing a convenient point as the origin, say the midpoint of the distance between the two fronto-zygomatic sutures, and using polar coordinates, the frontal view may be measured'.

This is illustrated in Fig. 5.1, reproduced from the paper:

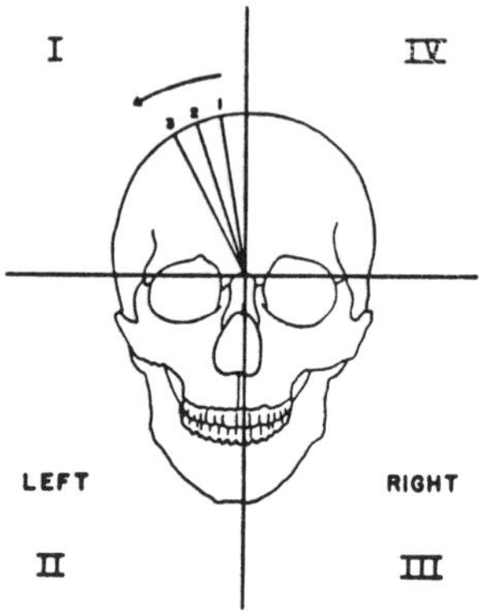

Fig. 5.1 Frontal view of a face, with the outline expressed in polar coordinates (Reproduced from K.H. Lu, *Biometrics,* 1965, **21,** pp.491-505, by permission of The Biometric Society)

'The measurements can then be plotted on linear scale over the range of $-\pi$ to $+\pi$. The graph in general assumes the form shown '.

This Figure is reproduced as Fig. 5.2. Lu then makes the new step: the value of the radius vector R as it is rotated about the origin is expressed as a function of rotation θ in a harmonic (truncated Fourier) series of the form:

$$R(\theta) = A_o + \sum_{n=1}^{m} B_n \cos n\theta + \sum_{n=1}^{m} C_n \sin n\theta \qquad \text{.................................... (5.1)}$$

It was in this form that Lu used the equation, but the form most generally used has been:

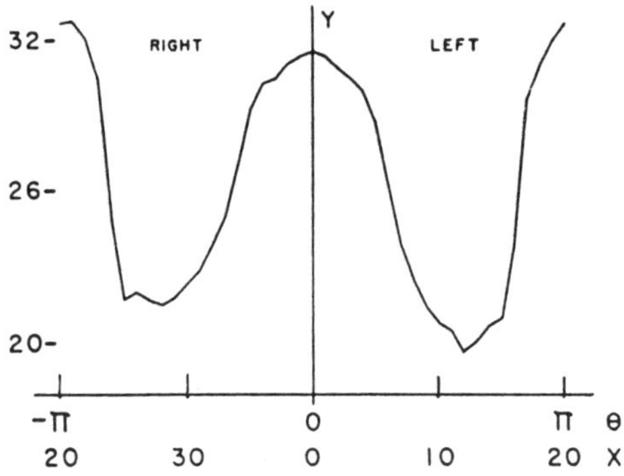

Fig. 5.2 Linear plot of polar coordinates (Reproduced from K.H. Lu, *Biometrics,* 1965, **21,** pp.491-505, by permission of The Biometric Society)

$$R(\theta) = A_o + \sum_{n=1}^{m} A_n \cos\,(n\theta - \phi_n) \qquad \text{.....................................(5.2)}$$

where ϕ_n is the phase angle of the nth harmonic given by $\tan^{-1}(C_n/B_n)$ and $A_n = (B_n^2 + C_n^2)^{1/2}$. The effect due to size variation is reduced by normalising the Fourier coefficients by dividing them by A_o, the mean value of all the R values. Both the coefficients B_n and C_n were used by Lu in his research. With this approach, he was able to study the changes in the outline shapes of boys' and girls' faces with age. Lu used a three-cycle Fourier equation and acknowledges the assistance of Bliss (1958), who used a similar three-cycle equation to fit some iodine values of butter-fat in milk recorded over time and expressed in terms of time.

It is interesting that Whittaker & Robinson (1924) state:

'In many cases the harmonics above the third may be neglected', and go on to show how, in this case, the coefficients may be calculated by 'simple averaging of the data without any multiplications'.

The pioneering work by Lu on the closed curve of a face outline does not appear to have been followed up, but, at around the same time as Lu published his work, there were publications using Fourier analysis on open curves in geology. Preston & Henderson (1964) analysed oil-well data, Stone & Dugundji (1965) used Fourier analysis in mapping, Harbaugh & Preston (1966) modelled geologic features that vary approximately in a cyclic way, and Stabler (1968) analysed fold shapes. Another, even earlier, application was to rainfall. Horn & Bryson (1960) write:

'It can be shown that the precipitation curve of a station can be reproduced by adding six properly chosen sine curves possessing frequencies from one to six.

The best fit is defined as the best least squares fit. The sine curve is fitted to the observed curve so that the sum of the squares of the differences between the sine curve and the observed curve is a minimum.'

In view of what is to follow concerning the use of harmonic analysis to quantify particle-outline shape, the paper by Stone & Dugundjii is of some interest. The authors state that the main problem is two-fold: (i) to present a precise, clear-cut definition of what is meant by the 'roughness' of a given curve or terrain, (ii) to devise some numerical expression for 'roughness' based on the definition adopted in (i). Large-scale topographic maps were prepared and a series of radial profiles were constructed through each of the areas at intervals of 15°. The curves were then subjected to Fourier analysis to give a series of sine curves, which, added together, gave a curve approximating to that of the terrain.

Stone & Dugundjii go on to discuss these observations:

(a) of two sine waves having the same wavelength, the larger one would be called 'rougher' by most people,
(b) of two sine waves having the same amplitude, the one with the shorter wavelength would be called 'rougher',
(c) the profile of a washboard terrain is usually conceded to be smoother than that of one having the same number of ridges, but irregularly spaced.

These quantities are independent, and a terrain can be 'rougher' in one respect, but not in another. Thus 'roughness', the authors state, is composed of at least the three independent factors listed; and, if it is desired to express the 'roughness' of a given terrain, it is necessary to specify the extent to which the curve possesses each of these three factors. It is the components of a 'roughness vector' that serve to give a picture of the nature and kind of roughness that the terrain curve has.

On an even larger scale, harmonic analysis had already been used on terrestrial topography. The calculations were, in fact, made on closed curves: the topography was analysed along parallels of the Earth. The amplitude and phase of the first 15 harmonics were presented (Peixoto et al., 1964.)

To return to the outline shape of a particle, it is not immediately obvious why the linear plot of the polar coordinates should be analysed by harmonic analysis. The fitting of curves to a set of data points is an exercise that is frequently undertaken. An example, reproduced from Duda & Hart (1973) in Fig. 5.3. shows five points with certain curves.

The straight line fits the data fairly well, the parabola provides a better fit and the tenth-degree polynomial fits the given data perfectly. Any function defined at a finite number of points can be represented by a polynomial if enough terms are taken. A technique in the science of statistics is to examine experimental data for significant trends by orthogonal polynomials. Each coefficient in the regression is

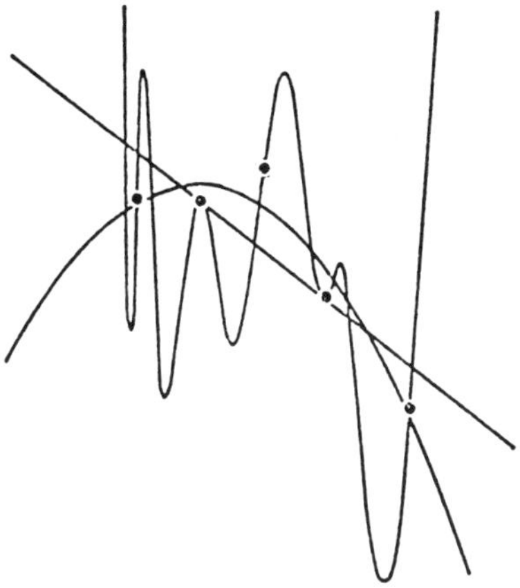

Fig. 5.3 Fitting curves to a set of data points (Reproduced from R. O. Duda & P. E. Hart, 'Pattern Classification and Scene Analysis'. Wiley-Interscience, N.Y., Copyright ©, 1973, John Wiley & Sons Inc., by permission of the publishers)

found independently of the others without the labour of solving simultaneous equations. This is a powerful tool, but it is not suitable for the reduction of the polar coordinates generated by the rotating vector, for three reasons:

(i) although the polynomial can be computed to pass through every one of the experimental points, the curve may be extremely badly behaved between points,
(ii) a small change in the orientation of the particle outline will cause the generation of a completely new set of coefficients,
(iii) polynomials are not periodic, and, with particle outlines, a describing function needs to have the same value at $\theta = 0$ as at $\theta = 2\pi$ for example.

Harbaugh & Preston, mentioned above regarding their geological data, state that, if the Fourier series provides a good representation of observed data, one would suspect that the variations are due to causes that vary in some periodic manner. Conversely, they suggest, if the series does not represent the data convincingly, other means of representation might be sought, as for example, power-series polynomials. This is an example where it is inaccurate to use the expression 'Fourier series' when it is being truncated. If sufficient terms are taken, a good representation of observed data will be achieved.

A good example of how few terms can be used with biological material is in the paper by Kaesler & Waters (1972). These authors used the rotating-vector method to study a particular fossil shape and required only a five-term series. This study is an interesting example of the quite sudden use of the harmonic-series approach; Benson (1967) had previously turned the fossil outline into polar coordinates with respect to a point within the outline and obtained the curve of perimeter radial lengths against the angle of rotation of the radials, but had gone no further than this so-called 'unrolling'. Jarvis (1976) considers that the harmonic-series analysis 'is a straightforward extension of Benson's theta-rho analysis.'

There must be a question as to why there should be almost a step-change from single-number outline-shape description to multinumber, series representation. The answer has to be the introduction of high-speed computing power. Reference to classic mathematical texts shows the calculation necessary for just 12 pairs of values and even more for 24 pairs. Large, convenient forms are reproduced in the text by Whittaker & Robinson from those used in the Mathematical Laboratory of the University of Edinburgh to assist with keeping the calculations under control in logical order.

Chernoff writes of his representation by faces:

'One is led to ask two questions. First, if this simple idea is so good, why wasn't it thought of before? Second, what alternative representations are there for points in high-dimensional space?

Introspection would suggest that this idea must have been considered before in simpler form. However, the effective application in this form would require a computer technology, which has only recently become available.'

An answer to the second question had been published the year before (Andrews, 1972). The proposed method of plotting high-dimensional data is suitable for inclusion here because, if the data are k-dimensional, each point $x' = (x_1, \ldots\ldots, x_k)$ defines a function:

$$f_x(t) = x_1/2^{½} + x_2\sin t + x_3\cos t + x_4\sin 2t + x_5\cos 2t + \ldots.$$

This function, which is a form of eqn. (5.1), is then plotted over the range $-\pi < t < \pi$. A set of points appear as a set of lines drawn across the plot. This is illustrated in Fig. 5.4. taken from the paper by Andrews. The means of eight measurements on several examples of a particular tooth from several species are involved.

It will be noted that the humans and the apes form fairly distinct groups and that the chimpanzees stand out from the gorillas and orang-outangs.

The idea of using a mathematical series to represent the numbers delineating a particle outline is an attractive one. In fact the use of two series has been reported in the literature: Fourier & Walsh, with passing reference to the possibility of using others, such as Haar (Meloy, 1977a). The difficulties mentioned above do not arise, and the series are orthogonol. These series will be dealt with in some detail, but it should be noted that there has been much mathematical activity on the subject published in the shape literature. This literature will be noted, but detail must be left in the originals and in standard mathematical texts. Good examples of such texts are those on the Fourier functions by Champeney (1973, 1985, 1987).

There has, of course, been a long history of the application of Fourier analysis to

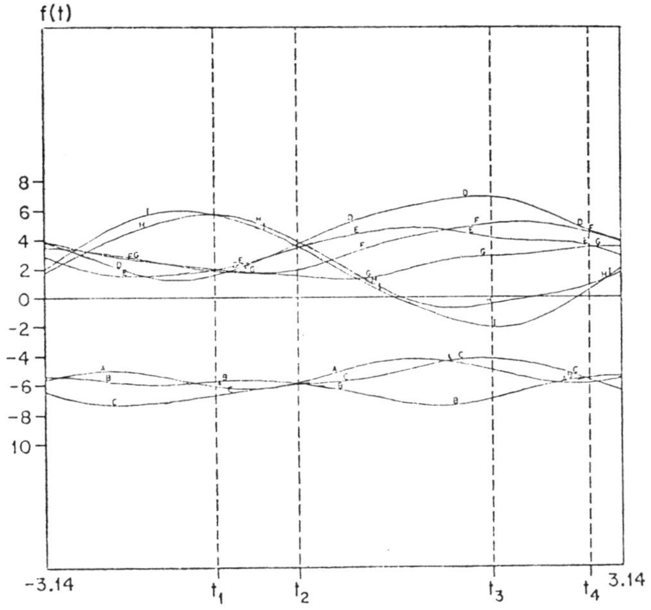

Fig. 5.4 8-dimensional data for permanent first lower premolar. Group means for: A = West African; B = British; C = Australian; D,E = Gorilla; F,G =Orang-Outang; H, I = Chimpanzee.
$f(t) = x_1/2^{1/2} + x_2\sin(t) + x_3\cos(t)$ ....
(Reproduced from D. F. Andrews, *Biometrics*, 1972, **28**, pp.125-136, by permission of The Biometric Society)

experimental data collected in, for example, physics, meteorology and biology since J.B. Fourier published his 'La Théorie analytique de la chaleur' in 1822. However, there is an important difference between those applications and most, if not all, of those concerned with particle-outline shape. The difference is that in conventional analysis regular oscillations are being sought, whereas in the analysis of outline shape the curve being analysed is normally nearer to being a random one. In the first case, the analysis provides the harmonics that make up the curve. Tests of significance have been available for many years (Hartley, 1949). In the second case, the harmonics are calculated that describe the curve. Bliss, for example, was searching for an underlying sinusoidal curve, whereas Lu was seeking an approximation for his unrolled curve. In other words, in the second case, regular oscillations are being imposed as the make-up of the curve. What is more, although the terms are orthogonal, it has been reported that, because of particular features of an outline requiring particular terms to be strongly present in a series, statistically significant correlation between terms can occur (Ehrlich et al., 1980).

The principle of building up the harmonic series is shown in Fig. 5.5 with a set of polar diagrams. ($R = 9 + \cos 5\theta$ was used by Sir D'Arcy Thompson as a diagram to illustrate a corolla of five petals.)

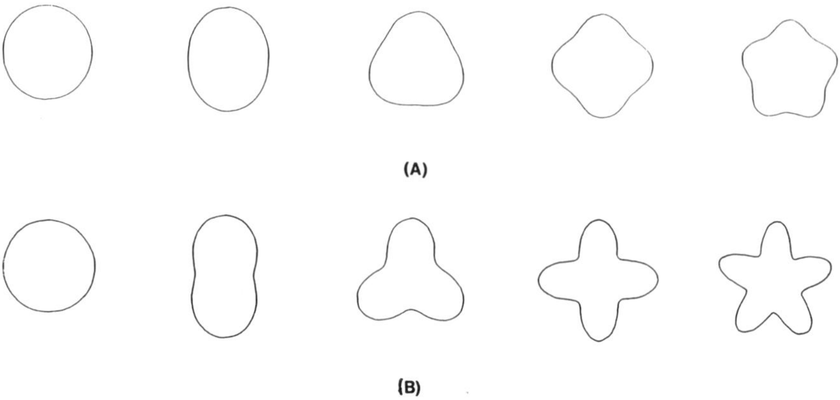

Fig. 5.5 Two sets of polar diagrams to illustrate some of the terms that are used in harmonic analysis
(A): R=9+cos$n\theta$, n=1,.......5; (B):R=3+cos$n\theta$, n=1,.....,5)

When a truncated Fourier series is used, these polar diagrams are effectively manipulated about their centres in amplitude and in angular position (i.e. phase) until the sums of the radii at all relevant angles are approximately those of the outline shape concerned. The more known points there are available around the periphery and thus the more series terms calculated and used, the closer the approximation. The general shape of an outline can be assessed from the first few terms.

The fact that certain early terms in the series indicate the general shape is, in a way, reiterated in a paper concerned with crystallographic problems (Rösler, 1987). Axial symmetrical outline forms such as those that exist in crystallographical research (squares, octagons etc.) were used to generate the appropriate harmonic series and the relation between the shapes, and the dominant Fourier coefficients was examined.

Walsh functions can be used to approximate to other functions in a manner similar to that in which sine and cosine functions are used in Fourier analysis. The first eight orthogonal Walsh functions are given in Fig. 5.6. Fox & Brazee (1971) remark that:

'The Walsh elements are not a function of frequency, as are the sine and cosine functions. To provide a basis for discussing the Walsh (and some other orthogonal functions) elements, a generalization of frequency, called sequency, is often used. Sequency is defined as one-half of the average number of zero crossings in an interval. Thus for a sine or cosine function, sequency and frequency are identical, but sequency can also be used to describe the relative spacing between positive and negative peaks of aperiodic functions.'

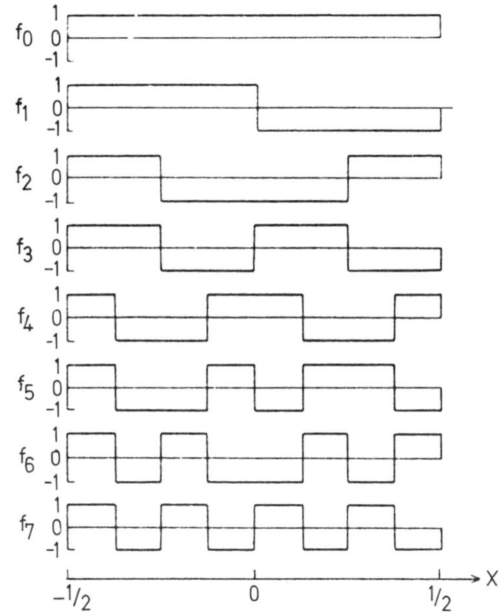

Fig. 5.6 The first eight orthogonal Walsh functions (Reproduced from R. D. Fox & R. D. Brazee, *J. Agric. Eng. Res.*, 1971, **16,** pp.353-363, by permission)

Walsh functions appear to have been introduced into the quantified study of leaf shape by Meltzer et al. (1967). The work was extended by Fox & Brazee, but both publications use the leaf in a frame as indicated in Fig. 5.7 (A), with the Walsh functions extended to two dimensions to give the Walsh patterns as indicated in Fig. 5.7 (B).

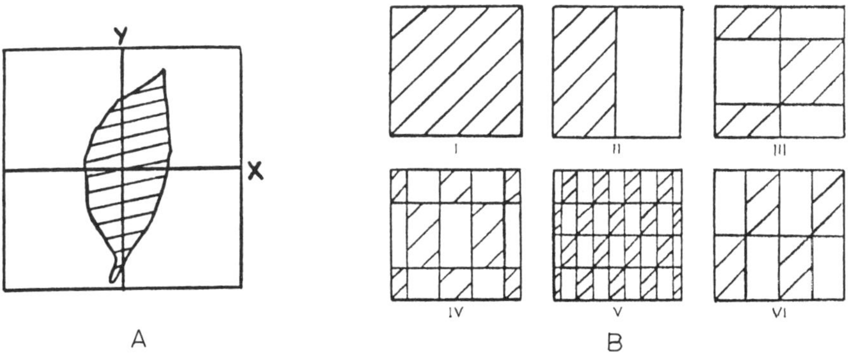

Fig. 5.7 A = object in frame as function; B=six basic Walsh patterns (Reproduced from B. Meltzer et al., (1967) Nature, **216,** pp.32-36, Copyright 1967 Macmillan Magazines Ltd., by permission of the publishers)

This would be a difficult procedure to use with particle outlines; both teams had to align all the leaf shapes, but still had problems with the data generated. If terms from a Walsh series are plotted in polar coordinates, it is clear that a particle outline would need to be of a ragged nature for a reasonable representation to be made using such a series. In Fig. 5.8, the two outlines have amplitudes comparable to the triforms in Fig. 5.5.

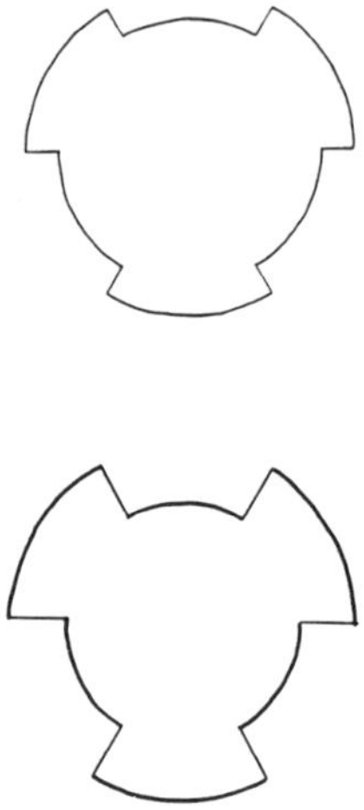

Fig. 5.8 Two polar diagrams obtained with Walsh functions for comparison with $R = 9 + \cos 3\theta$ and $R = 3 + \cos 3\theta$ shown in Fig. 5.5

In the paper in which Meloy introduced the Walsh functions into powder technology it is indicated how the functions might be used; this was done by a comparison with Fourier functions, both plotted in polar coordinates. Beddow (1980) reproduces some of the Meloy figures and suggests that the advantage of representing particle profiles by square-wave functions instead of by Fourier functions may be understood from the figures. In addition to the elliptical and triangular shapes being preserved by the Walsh representation, it is possible to calculate immediately the excess arc length for the profile, as represented, over the base-circle radius A_0. This is not possible with the Fourier representation, but the question arises as to the value of this calculation when it is for a representation of a two-dimensional outline of a particle. It would perhaps be unwise to convert this perimeter difference into a difference in surface area.

The use of the Walsh series has not progressed in particle-outline measurement. This may have been due to unfamiliarity with square-wave functions, but is more likely to be due to the restricted shape representation that results. As Flook (1982) points out ,'most particle waveforms do not exhibit sharp discontinuities and were therefore better suited to synthesis by circular waveform functions'.

Examination of the figures in the paper by Meloy shows that the Walsh functions are slow to develop detail, and Flook states in the discussion of his paper:

'It would require approximately two or three times the number of Walsh terms than would be required for a Fourier series expansion to give the same degree of accuracy. Although the binary nature of Walsh functions gave them a speed advantage in their computation as compared with Fourier algorithms, the primary objective of shape analysis must be to restrict the number of coefficients required and therefore the author saw little advantage in their use for this purpose.'

In the light of research being parochial, a paper published two years after these remarks by Flook is of interest. Sergeev et al. (1984) give a full description of the truncated Fourier series for the morphological analysis of scanning electron micrographs of quartz particles, referring back only to the original paper by Ehrlich & Weinberg (1970). In addition to calculating the same kind of 'roughness' terms as Ehrlich & Weinberg, the authors calculate the area and perimeter of the particle outlines and calculate a 'shape complexity coefficient *K*', given by:

$$K = 4\pi \text{ (area)} / \text{(perimeter)}^2$$

This is the 'roundness' index defined by Cox in 1927. The paper by Sergeev et al. is important, however, because it describes in detail the use of Walsh coefficients to characterise the quartz-particle outlines, although no mention is made of any similar work such as that by Meloy. The authors found that, to describe an outline shape adequately, more Walsh coefficients were required than Fourier, but that the time taken to calculate ten of the latter was sufficient to calculate 600 of the former.

The outlines of cities used by Boyce & Clark (1964) and referred to in the previous Chapter would be suitable for description by Walsh functions because of their square or rectangular shape, but this would seem to be exceptional and the Walsh functions will not be discussed further.

The application of Fourier analysis to particle outlines has been identified by Meloy (1983) with three schools: Meloy publishing in 1970, Ehrlich publishing in 1970 and Beddow publishing in 1975. To these names of early workers should be added Cheng & Sutton (1971) and Sutton & Bundalli (1973); their papers have not had the same influence as the others and are rarely cited, which is a pity as the accounts are particularly clear and succinct.

The first published paper to describe the use of harmonic analysis to quantify the outline shape of grains seems to have been that by Schwarcz & Shane (1969), which is sometimes given as a reference but had nowhere near the influence of the paper by Ehrlich & Weinberg (1970). There is no reference in this latter paper to the earlier Schwarcz & Shane paper, which described similar ideas published in a journal from the same sedimentology literature; the latter was, in fact, submitted before the former!

Schwarcz & Shane relate their harmonic analysis back to the traditional ideas of 'roundness' and 'sphericity'; Ehrlich & Weinberg do not. It is important to note that almost all the papers referred to in this Chapter have something to add to the technique; it is advisable to study several of them before adopting this method of outline shape quantification.

The numerical method used by Schwarcz & Shane to calculate the Fourier coefficients requires that the outline be sampled at points of equal angular separation. This was achieved by using a transparent polar-coordinate grid. The origin of the coordinate system was found in a manner different from that used by other, later workers. An initial origin was chosen at the centre of the smallest circumscribed circle and a mean-square procedure applied to the distances from that centre to points on the outline to obtain a 'centre' of the grain. This final centre was used to give the points on the outline used in the subsequent shape analysis. Schwarcz & Shane state that the centroid of the image could have been used as the origin, and it will be found that that is what nearly all later workers have used, except some of those working with biological material, e.g. molluscs, where a natural reference point exists. Kaesler & Waters (1972) have noted that a high variation in the first harmonic can arise from the use of some natural reference point as the origin because of biological diversity.

Schwarcz & Shane measured at 5° intervals, and normalised their values to avoid the shape parameters depending on the size of the measured image. They write:

'For the purposes of numerical solution we limit ourselves to a finite number of terms in the trigonometric series. Strictly speaking such a series is not a Fourier series since it is finite. However we shall continue to refer to it as such in the remainder of this paper.'

Three different measures of roundness, calculated from the amplitudes of harmonics, were proposed and shown to correlate with visually estimated roundness classes, but this does not seem to be a promising use of the detail available with harmonic analysis and does not appear to have been pursued elsewhere except by Huller & Weichert (1982). In the discussion of this paper, it was suggested 'that most of the information on particle shape contained in the Fourier analysis was lost when the coefficients were combined to give a shape measure such as roundness.'

The approach of Ehrlich & Weinberg was one of identification or 'fingerprinting' and provided a useful way forward. The authors write:

'few, if any, researchers have successfully used shape statistics as a basic parameter'.

After their paper outlining the Fourier method was published, there was a flood of apparently successful work in geology. Hawkins (1980) lists 18 references, and there have been many more since then. Several of the papers include work by the Ehrlich team on problems that are analogous to those facing powder technologists because the investigations are into the sources of the examined material.

The manner in which the calculations are performed does depend, to a certain extent, on the way in which the original data points were collected. One method has already been described: overlaying a radial grid and collecting up points at the required intervals of θ. This is suitable for manual or semiautomatic operation, but can be subject to operator error if care is not taken. The method used by Ehrlich & Weinberg involved the recording of rectangular coordinates of points around the periphery of the outline and from them the calculation of the centroid of the area. Each recorded point on the periphery may now be expressed in polar coordinates about the centre of gravity. As Clark & Clark point out (1976) :

' ....it is easier to solve a Fourier series if the data points are equally spaced (in this example, at equal angular separation). Ehrlich & Weinberg use a linear interpolation scheme to provide the equal spacing, but this could introduce unwanted bias. Here equal spacing was achieved by first fitting a bicubic spline (Ahlberg et al., 1967) to the grains periphery. A spline has the property of passing through all the data points as a smooth curve. New data points at equal angular increments were calculated on the basis of the spline.'

The above procedure is illustrated by an outline shape digitised at only 34 points. In a later paper (1981), Clark remarks apropos of the desirability of having equally spaced data and the fact that the points collected for the calculation of the centroid are unlikely to bear the desired relationship to the centroid:

'Either the vertices must be collected again (a strategy which appeals to no-one) or the derived values must be interpolated from the observed points. In fact, hidden well within Ehrlich & Weinberg's equations is a linear interpolation to obtain these values. There is also an error in that interpolation, pointed out by Jarvis (1976).'

The paper by Jarvis, on drainage-basin shape, has been referred to before in this book. The error referred to is that the shape being analysed is made up of points joined by straight lines and the interpolation should be made along the straight lines. This is not the case in the treatment by Ehrlich & Weinberg, and the difference might be significant if only 34 points around the outline were used, but it is small with an adequate number. The same argument applies to the use of a spline rather than straight lines.

In a more recent paper, Full & Ehrlich (1982) describe how the quality of the harmonic series derived for an outline shape may be improved. This is accomplished 'by minimising the distance between the origin of the polar coordinate system in the calculation of the Fourier series and the shape centroid'. The contribution is an important one. It deals with the problem of computing the centroid of relatively complex outline shapes.

Gevirtz (1978) computed the centroid of bivalve outlines from 72 data points spaced at equiangular intervals with a grid on an arbitrary centre. From the computed centroid, new radii were extended to the same data points. This is a useful paper because the author demonstrates the importance of significant harmonics and how intraspecific and interspecific variations can be examined. In the section on computation, Gevirtz writes:

'Fourier analysis provides an exact description only for measured data: because curve sections occurring between data points can only be estimated, sufficient measurements must be taken to reasonably account for all visible variation in the original form.'

It is essential that there be sufficient points for the polygon to represent adequately the shape under study; otherwise, the centroid and all else will be computed for the wrong shape. Rather smooth, round outlines such as that presented by Clark & Clark mentioned earlier can, perhaps, be dealt with by 34 points, but a shape such as that given by Jarvis, reproduced in Fig. 5.9, cannot.

For any particular type of outline, it is possible to establish how many terms should be used, up to the Nyquist limit (Davis, 1973), by generating the data using more and more coordinates until near-constancy is achieved. If manual digitisation is being used, it is possible with pre-planning to employ the digitising procedure with a higher efficiency by using many points over a length where the shape is complex and fewer points where the outline is approximately straight.

Some measure of goodness of fit is needed. Lu points out that 'the observed values are (except for small errors of measurement) the true values, while the calculated values are only approximations to the truth, resulting from our efforts to obtain an effective representation; neither can properly be regarded as a random variable. There is thus no question of a true significance test; we are concerned merely to obtain some measure of the closeness of agreement'. Lu puts forward the empirical requirement that there is a 'correlation' between the two values of about 0.9. As a test of convergence, Schwarcz & Shane state, 'the original $r(\theta)$ values were compared with values calculated using the Fourier series'. For 35 terms, the agreement was within one part in 10^{-4}. Ehrlich &Weinberg in their paper do not concern themselves with this problem, but, like many authors, demonstrate how an outline shape is reproduced with an increasing number of terms. This is shown in Fig. 5.10, from their paper. Centres of the small circles surrounding the Figure are said to represent actual points on the periphery of the original grain. By today's standards, unfortunately, this pioneering work appears unpolished. The ways in which the work has progressed are possibly best seen in a paper by Kennedy (1985). This reviews papers by Ehrlich and some of his collaborators, touching on harmonic-amplitude structure, multivariate factor and cluster analysis.

A Fourier or harmonic series can expand only a single-valued function; so it is important to note that a radius drawn from the centroid of an outline must intersect only one peripheral point. An example of an unsuitable outline was illustrated in Fig. 3.6 in Chapter 3.

It is clear that, with hand digitisation, if this problem should occur it can often be taken out, but with automatic image analysis it is possibly best to have any such

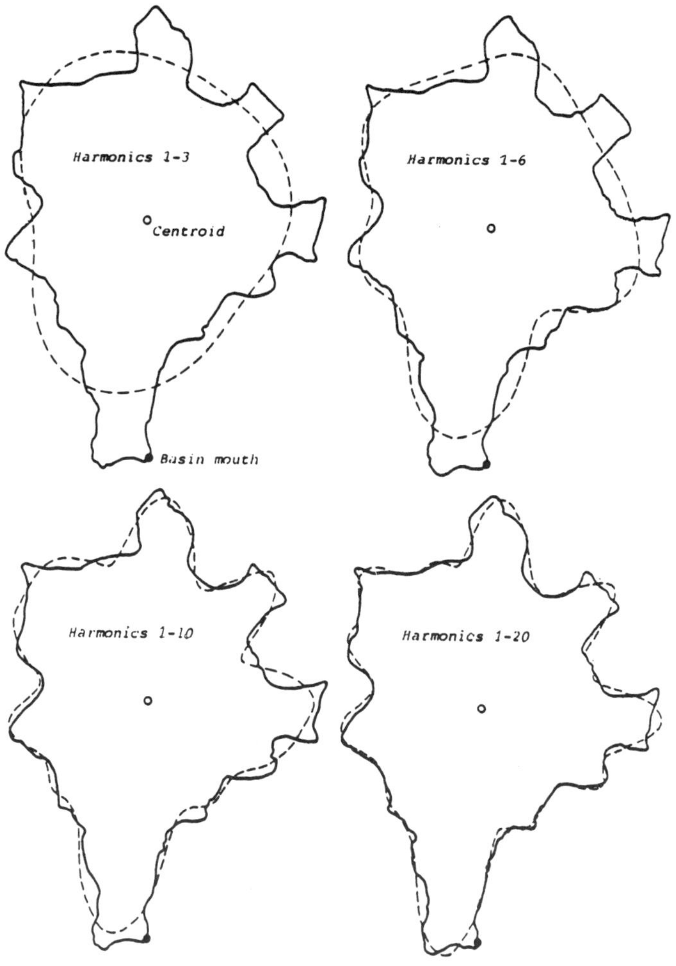

Fig. 5.9 Modelling the shape of the River Ottery basin. The observed perimeter is shown by the solid line, and the generated shape by the dashed line (Reproduced from R. S. Jarvis, *Water Resources Res.*, 1976, **12,** pp.1151-1164, Copyright by the American Geophysical Union, by permission)

outline rejected by the computer program. However, although there has been much mention of this matter in the literature, others have not found that it causes significant problems. Indeed, at least one early paper states that this kind of problem is extremely rare. However, with manmade materials, a generalisation is not advisable, although Holt (1981) in a fresh approach to crushing and milling found that only a few particles had re-entrant features.

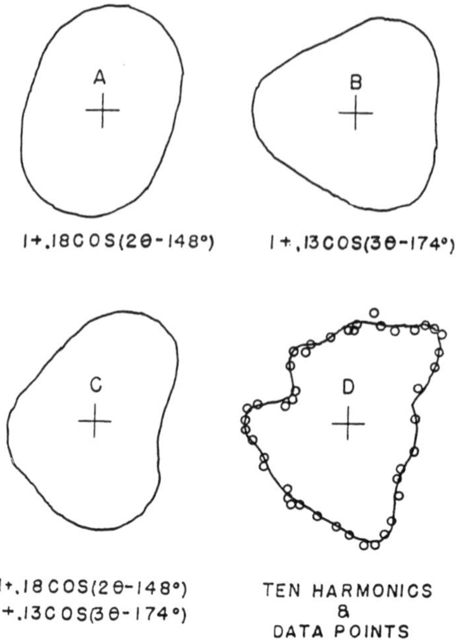

Fig. 5.10 Harmonic contributions to shape (Reproduced from R. Ehrlich & B. Weinberg, *J. Sedim. Petrol.*, 1970, **40,** pp.205-212, by permission of the Society for Sedimentary Geology)

A similar harmonic-analysis approach avoids this, so-called, re-entrant problem. The particle outline is represented by an Angle Versus Length (AVL) representation. This is a technique that, from the boundary, generates a waveform for later harmonic analysis by plotting the tangent angle versus arc length around the boundary so as to form a closed contour. This is illustrated in Fig. 5.11. Fong, Beddow & Vetter (1979a) were the first to introduce the scheme into powder technology. The authors state that 'The method described in this paper utilizes a procedure first described by Zahn and Roskies (1972)'. In fact, an extensive paper on the method by Brill (1968) states that the AVL representation and Fourier descriptors were conceived by Cosgriff (1960) and subsequently refined and developed by others. In the course of the paper, Brill mentions that two of the workers had investigated several mathematical expansions and concluded that the Fourier-series expansion held the most promise. Brill outlines the application of AVL thus:

'Although the paper limits itself to character recognition, the techniques described herein are applicable to any class of pattern recognition problems in which the

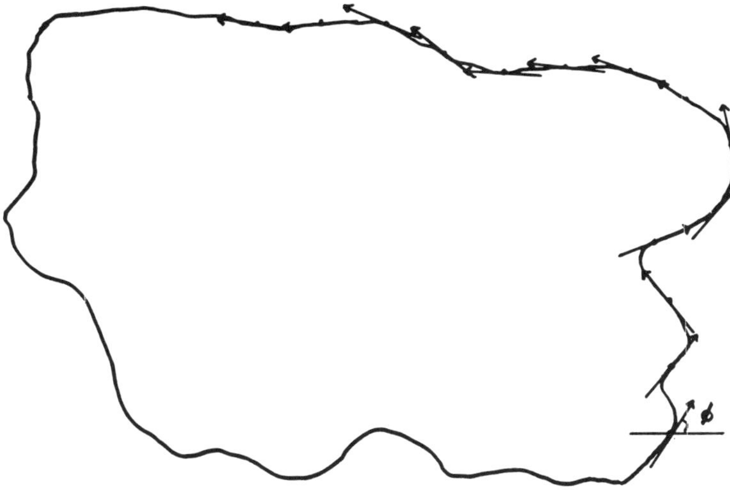

Fig. 5.11 An outline parameterised by its arc length and change of slope

objects to be identified can be classified on the basis of their boundary edges. These methods have been applied to automatic recognition of targets on aerial maps, aircraft recognition, identification of ground features from contour maps, identification of archaeological fossils such as sea shells, and fingerprint identification.

In addition, the Fourier descriptors permit a qualitative comparison of character shapes and provide an indication of the presence or absence of symmetry. For synthesis problems, reconstruction of the pattern by combining different subsets of its harmonics expose (sic) properties that may be useful for better character perception, compact information transmission, etc.'

Despite the practical use of this method of outline representation (Fong et al., 1979b; Lee et al., 1980), a serious difficulty is hidden in the accounts. If reference is made back to the rotating-vector method and, in particular, to Figs. 5.9 and 5.10, it will be seen that an outline shape may be represented crudely by a very few terms in the harmonic series or, if required, better represented by more terms. This fact is further illustrated in Fig. 5.12, taken from the paper by Hawkins (1980). Now, if reference is made to the paper by Fong, Beddow & Vetter, it will be found that the illustrations show regenerate shapes using 75 coefficients; two examples are reproduced in Fig. 5.13. This is significant. Jones (1983) points out succinctly and graphically that a sensibly truncated Fourier series cannot regenerate outlines by

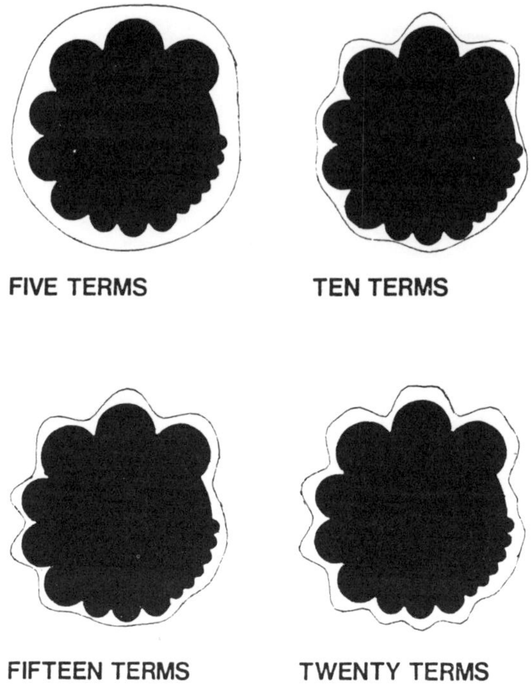

Fig. 5.12 An artificial shape regenerated using 5, 10, 15, and 20 terms derived from 180 equally spaced points (Reproduced from A. E. Hawkins, Proc. Technical Sessions, Powder Europa, Wiesbaden, 1980)

the AVL method; the method suffers from a problem of closure. Jones takes the outline of England, Scotland and Wales to show how, by using the harmonic series, the approximation of the tangent angle at each step of the reconstruction leads the polygon away from its true finishing point. The approximations are inevitable as a result of using only a few terms of the series. In his short paper, Jones states that the 'problem can be overcome by recognising that the regenerate curve generates the shape in a clockwise sense; if we also generate the shape in an anticlockwise sense, it is possible to meld the two curves into one single, closed curve'. To illustrate this, reproductions from the paper are made in Fig. 5.14. This is a serious difficulty because, although Meloy (1977a) was happy to contemplate 1000 Fourier coefficients for the rotating-vector method, most papers refer to far more modest numbers and, even then, are concerned about the quantity of data. To a certain extent, since the maximum number of Fourier coefficients is determined by the number of coordinate pairs, manual digitisation keeps the coefficient numbers down whereas automatic image analysis, if time is no object, can provide hundreds.

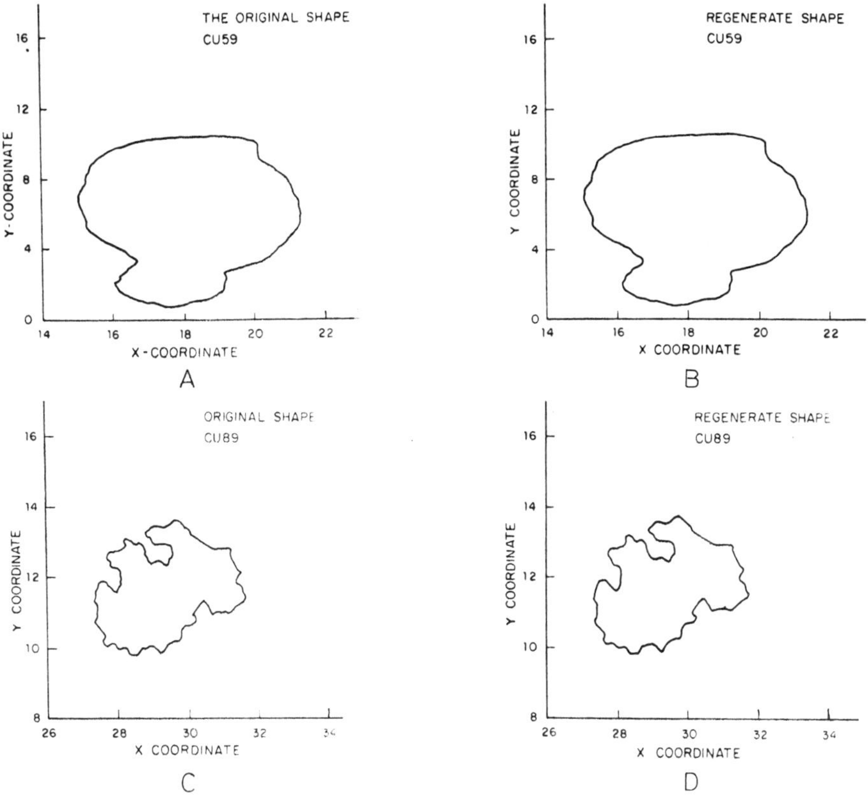

Fig. 5.13 Original shapes and regenerates using 75 coefficients from the harmonic series derived by the angle-versus-length representation.
A, C = original shapes; B,D = regenerate shapes (Reproduced from Fong, S. –T. et al., *Powder Technol.*, 1979b, **23,** pp.187-190, by permission of the authors and Elsevier Sequoia S.A.)

Gotoh (1979) picked up this problem:

'The change in direction is apt to deform the original shape owing to the accumulated error. The ordinary polar coordinate is most accessible. But it fails in expressing non-holomorphic forms where the radial distance $r = r(\theta)$ becomes multivalued.'

The AVL method has been applied to at least one practical research activity with some success. The silhouettes of copper particles of three types (gas atomised, reduced, and the complex and re-entrant electrolytic) were quantified (Singh & Ramakrishnau, 1988). For analysing the Fourier data, Singh & Ramakrishnau used mean-harmonic-amplitude spectra, frequency-distribution histograms and chi-square methods, but they had to compute 50 harmonics.

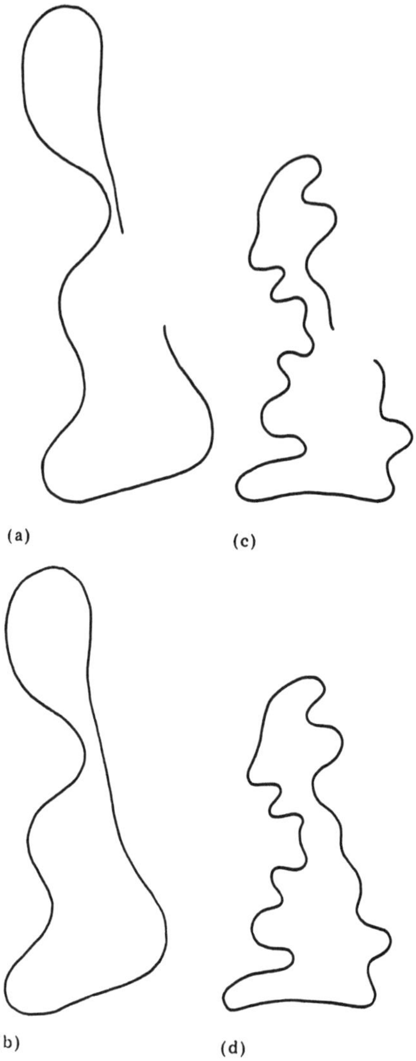

Fig. 5.14 A map of Britain regenerated from the Zahn & Roskies reconstruction formula and from a closed regenerate formula: (a) truncated level 6 – unclosed form; (b) truncated level 6 – closed form; (c) truncated level 20 – unclosed form; (d) truncated level 20 – closed form (Reproduced from S. Jones, *Powder Technol.*, 1983, **34,** pp.93-94, by permission of the author and Elsevier Sequoia S.A.)

Réti & Czinege (1989) have managed to combine the AVL method with the rotating-vector method in a useful clustering technique. The authors state that the combination evaluated morphological similarity well, because it included the phase information as well as the amplitudes.

To return to the rotating-vector method, Ehrlich & Weinberg in their creative paper plotted normalised amplitude coefficients against appropriate harmonic numbers as shown in Fig. 5.15. The relative positions allow subjective evaluation of the shapes, the Ottawa Sandstone having a relatively simple shape. Ehrlich & Weinberg went on to calculate a 'roughness coefficient', given by:

$$(\tfrac{1}{2}\sum_{n=1}^{m} A_n^2)^{\frac{1}{2}}$$

The symbols refer to the harmonic series given earlier in eqn. (5.2). In a review paper of their studies over ten years (Ehrlich et al., 1980a), the authors did not mention the roughness coefficient and concentrated on their powerful 'fingerprint' techniques:

'It will be demonstrated that the three major factors that influence particle shape – provenance, process history and transport rate – can often be clearly deciphered from shape data. This allows quartz grain shape to play the role that mineral composition usually plays in detrital sedimentology.'

It is necessary to make several quotations from this review paper and from a contemporary and similar book chapter (Ehrlich et al., 1980b) because of the practical demonstrations of the usefulness of the technique to sedimentologists and others and also to gainsay a depressing statement by Meloy(1986):

'In the early days of shape analysis, when the Fourier method was thought to be the answer to all shape problems, every powder was analyzed by Fourier, but this early optimism has been replaced by the somber realisation that few problems are solvable by Fourier.'

An important view of Ehrlich et al. is that the characteristics of a single grain are less important than the shape-frequency distribution of a sample of many grains. They write:

'Histograms were prepared to display the range of amplitude values *at a given harmonic for each sample*. These 'shape frequency histograms' where frequency of occurrence is plotted as a function of increasing harmonic amplitude, permit visual comparison between samples or between harmonics within one sample.'

An example is given in Fig. 5.16. Where there are significant differences, in this

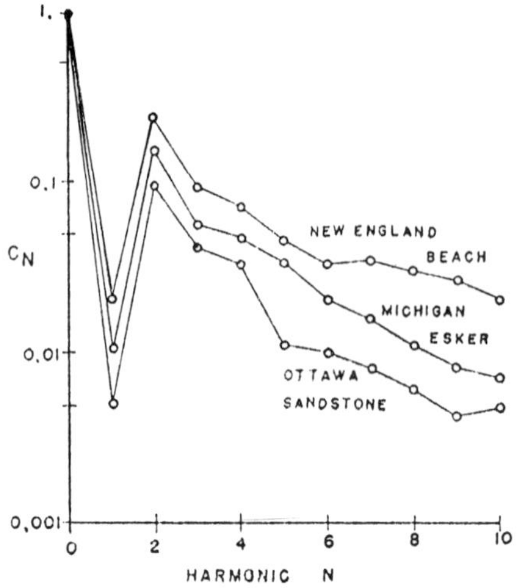

Fig. 5.15 Mean-amplitude spectra of quartz from three sands:
Michigan esker 26 grains
New England beach 14 grains
Ottawa Sandstone 33 grains
(Reproduced from R. Ehrlich & B. Weinberg, *J. Sedim. Petrol.*, 1970, **40,** pp.205-212 by permission of the Society for Sedimentary Geology)

case with the amplitudes of harmonics 6, 9, 14, 17, 19, 20, it should be possible to use these harmonics as 'fingerprints' to distinguish between samples from the two sources. A good example of such a use is the research into silting in Charleston Harbor, South Carolina (Van Niewenhuise et al., 1978). Put briefly, the question posed was whether the sand and silt within the area came from the rivers or the ocean. Using 20 harmonics, shape-frequency histograms were generated from 300 grains from each sample taken at the points shown in Fig. 5.17. For identification purposes, it was found easier to use frequency histograms which showed marked differences at their tails rather than within the body of the distribution. Chi-square results were used as a guide to visual inspection. Harmonic 14 was found particularly helpful, and the published histograms are reproduced in Fig. 5.18. The authors were able to find the sources of sediment influx into the harbour and also the relative contributions of each source; the potential for the use of the technique in powder-technology processing is obvious.

Some years earlier, Anstey & Delmet (1973), working with sections of fossils, also used plots of mean harmonic amplitude against harmonic number. They state:

'Although the mean harmonics of different taxa are distinctive, the variability (indicated by one standard deviation) shows strong overlap between different species and genera. Therefore, a measure of central tendency should be used for systematic differentiation, and not values obtained from only a few individual shapes.'

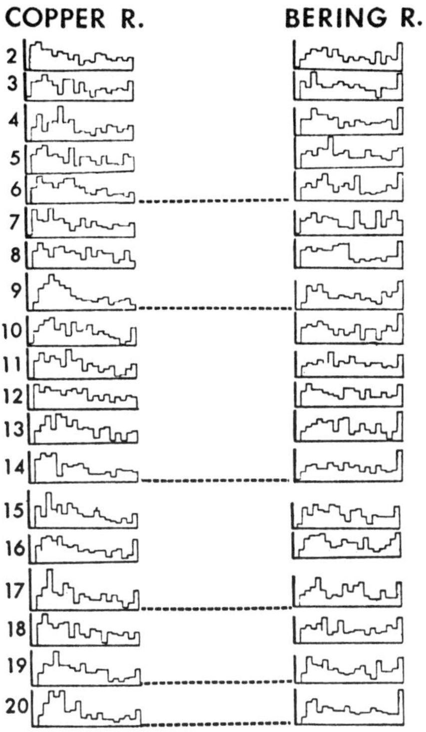

Fig. 5.16 Shape-frequency histograms computed for harmonics 2-20 for two samples from the Gulf of Alaska. On the horizontal axis, values denote class intervals of harmonic amplitude; the vertical axis represents the number of quartz grains per class interval. Dashed lines connect the histograms that showed significant differences at the 0.05 level using chi-square analysis (Reproduced from R. Ehrlich et al., *J. Sedim. Petrol.*, 1980a, **50**, pp.475-483, by permission of the Society for Sedimentary Geology)

In their book chapter, Ehrlich et al. write:

'The easiest part of our task has been the characterization of the shape of a discrete particle. The most difficult task has been to determine appropriate numerical-statistical techniques to extract information from the multivariate data field so generated. Because sustained and concentrated research in this area has not been conducted before, we could not utilize a large store of previously generated data, which forced us to begin our analysis with determination of the nature of the underlying frequency distribution. This, in turn, influenced the subsequent choice of numerical-statistical methods of analysis.'

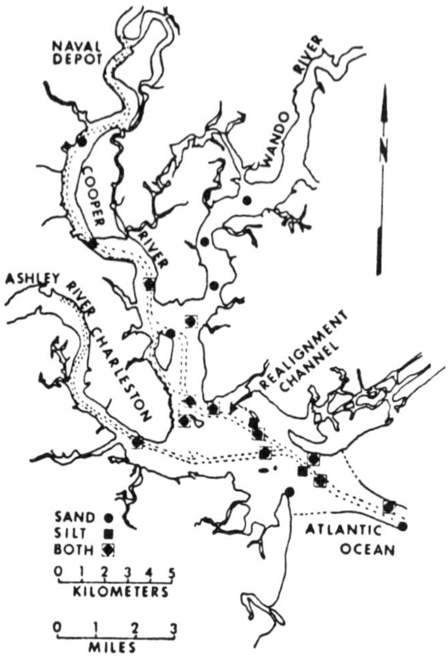

Fig. 5.17 Locations around Charleston Harbor where sand and silt samples were collected (Reproduced from D. S. Van Nieuwenhuise et al., *J. Sedim. Petrol.*, 1978, **48**, pp.373-383, by permission of the Society for Sedimentary Geology)

It will have been realised that Ehrlich et al. have used only part of the data in the harmonic series; the phase component was omitted. It should be noted that, if the phase angles are not required then, according to Meloy (1977(a)), it is possible to obtain the Fourier coefficients alone by the use of the Fast Fourier Transform. The advantage is given as speed of computation; if there are N equally spaced polar coordinates, this ratio applies:

$$\frac{\text{time to calculate by Fourier-series analysis}}{\text{time to calculate by Fast Fourier Transform}} = N/\ln N$$

Meloy (1977b) used the amplitudes alone in an original manner. A harmonic series is convergent, but Meloy suggests further that log-log plots of A_n versus n are linear to give a slope, an intercept and scatter about the line. Such a plot has been termed the 'signature' of a single particle, and the envelope of a set of such plots is termed the 'signature' of the powder made up of the particles. However, Austin et al. (1990) found that this particle-signature analysis did not distinguish between particle outlines of obviously different shape.

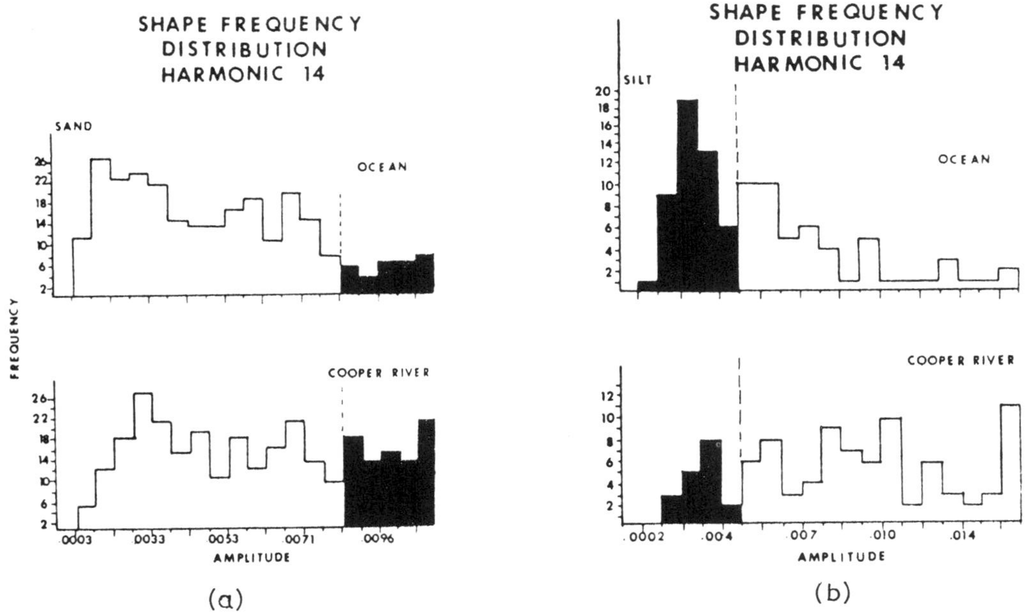

Fig. 5.18 Histograms of amplitudes of the 14th harmonic, illustrating differences between (a) fluvial and marine sands, (b) fluvial and marine silts (Reproduced from D. S. Van Nieuwenhuise et al., *J. Sedim. Petrol.*, 1978, **48**, pp.373-383, by permission of the Society for Sedimentary Geology)

There are, of course, standard texts written about the handling of multivariate data. In a particularly good example, Sir M. Kendall (1975) has suggested that the computer revolution has, perhaps, affected the development of multivariate analysis more than any other branch of statistics. Among the practical problems described is one in biometrics where a number of skulls are dug up on an ancient burial ground. They may all come from one race or may be a mixture of two opposing races, friend and foe flung into one pit together. An unlimited number of measurements can be taken on any one skull. What are the best measurements to take, what is the minimum number we require, and how do we use them to test homogeneity or heterogeneity in the sample ?

This is the kind of problem with which many of the workers referred to in this book have been faced. Among the many and various multivariate methods available is clustering. Given a set of objects and observations of their qualities, we require to form them into groups on the basis of their internal similarities; is there any evidence that they cluster into definable groups as against the alternative hypothesis that they are an unstructured set?

According to Kendall, the theory of cluster analysis was relatively new when he was writing, and the fact that there were many methods available might have been due to the need for precise and unambiguous data to be fed into a computer. Moreover, because the concept of 'cluster' was a subjective matter, different methods might be required to detect different patterns of points.

The general theme is that the extent to which two groups are to be regarded as separate clusters depends on the distance between their means as compared with the mean distance within the clusters. This is not a trivial computing problem; Kendall points out the distances between pairs rises alarmingly. With 1000 individuals, there result nearly 500 000 distances. Despite this, it is appropriate here to introduce a different approach to the problem of analysing the data from harmonic-series analysis that does use all the information within the shape equation. Both amplitudes and phases are used, and the outline shapes may be reproduced. The concept is not to 'fingerprint' particle outlines, but to cluster together similar outline or silhouette shapes.

The new approach to the problem of using the harmonic-series data in the most efficient manner was first put forward by Davies & Hawkins (1979). After noting that the amplitudes of the terms in an harmonic series are often used alone as shape descriptors, the authors continue:

'They cannot be used as the sole descriptors, however, because many shapes can be constructed which have the same amplitudes [Fig. 5.19]. This illustration shows three different shapes each with just the third and fourth harmonics present.

The difference between them is the relative orientation of the harmonics. In the left-most shape a third harmonic peak is coincident with a fourth harmonic peak. In the right-most shape we have the opposite extreme with a third and fourth harmonic

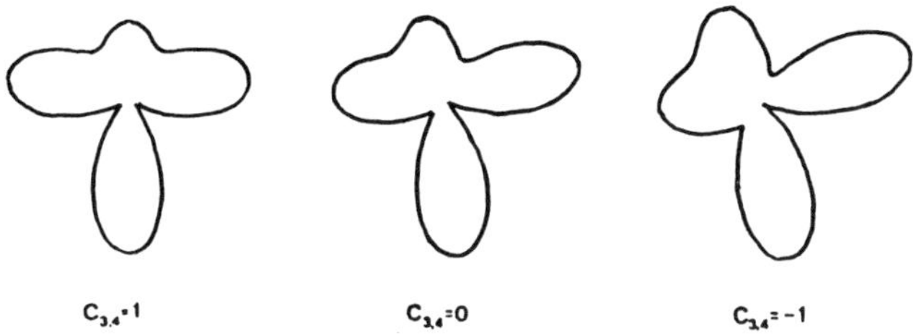

Fig. 5.19 Three outline shapes that all have A_0=1, A_3=0.5, A_4=0.5, but different phase relationships (Reproduced from K. W. Davies & A. E. Hawkins, Proc. 2nd European Symposium on Particle Characterisation, Nürnberg, 1979)

trough being coincident. The remaining shape shows an orientation mid-way between the two.

'A term which describes the relative orientation of two harmonics m and n is:

$$C_{m,n} = \cos\left[\frac{m\phi_n - n\phi_m}{\text{Greatest common divisor of } m \,\&\, n)}\right]$$

$C_{m,n}$ can range from +1 when harmonic peaks are coincident down to -1 when the peaks are at maximum separation. The emphasis we place on $C_{m,n}$ must be related to the importance of the m_{th} and n_{th} harmonics in the series. Thus we take:

$$P_{m,n} = (A_m A_n)^{1/2}.C_{m,n}$$

Each $P_{m,n}$ is independent of the origin and direction of rotation of the radius vector. This means that identical $P_{m,n}$ values will be obtained from a mirror image of the shape. The terms $P_{m,n}$ are symmetric, i.e.

$$P_{m,n} = P_{n,m}$$

and include the amplitudes, since $P_{m,m} = A_m$.'

It is these P values that are clustered.

In a test of methodology, the harmonic series for two sets of outlines were obtained and the appropriate P values calculated and clustered. One set of results is shown in Fig. 5.20. Each cluster contains shapes that are broadly similar; if the shapes are grouped with a large number of clusters, those in the same cluster will possess a high degree of similarity. The choice of the number of clusters to take will be considered later. The two sets of outlines were of flakes: one lot deposited on plain-glass microscope slides, the other lot deposited on slides to which double-sided adhesive tape had been attached. A total of 59 flakes on sticky tape were

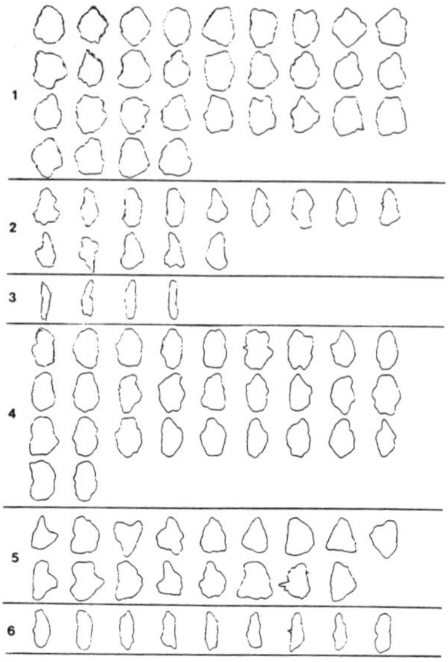

Fig. 5.20 Outline-shape clusters obtained from flakes scattered on microscope slides; one set of slides with double-sided adhesive tape attached, the other set of slides plain (Reproduced from K. W. Davies & A. E. Hawkins, Proc. 2nd European Symposium on Particle Characterisation, Nürnberg, 1979)

measured and 45 from glass. Thus, if there were no difference between shapes seen on the tape and on the glass, each cluster should split on average in the ratio 59:45. The actual splits are shown in Fig. 5.21. There was a statistically significant difference (Freeman & Halton, 1951), and it will be noticed that the narrow shapes (flakes on edge) were found on the sticky tape.

Davies & Hawkins present another example of this type of analysis where it was necessary to find out if any of four processes gave particle outlines similar in shape to those from a fifth process. In Fig. 5.22 the clusters that were formed are shown, typified by six particles from each cluster. The process origin of particles in each cluster is shown in Fig. 5.23.

The procedure just described can be applied generally whenever one powder must be compared with another. For example, the study of Charleston Harbor could have been made in this more thorough way. The fact that a great deal of computation is involved has been commented on in the literature (Flook, 1982a), but inexpensive computing power now available makes this a much less serious matter.

When discussing the number of harmonics to use to represent shape outlines, Davies & Hawkins consider some measures that might be useful:

'One such measure is obtained by considering the variance, W, of shapes within a cluster compared to the total variance, T, of shapes overall. A measure of clustering

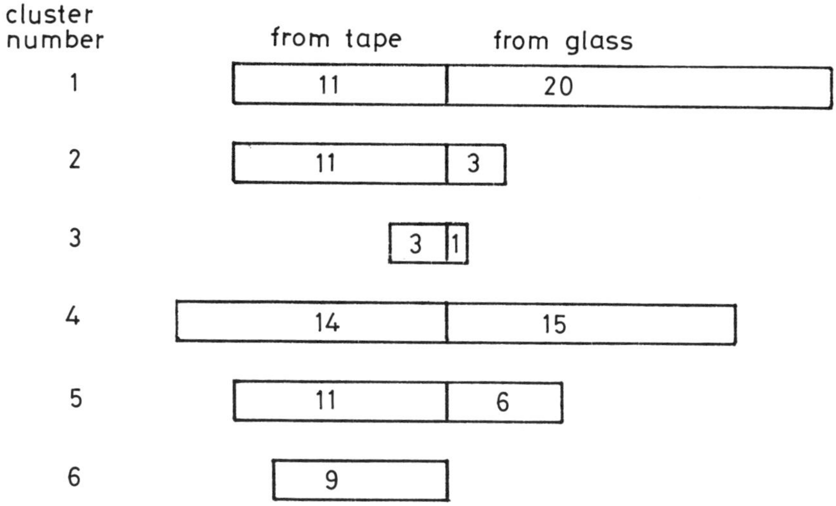

Fig. 5.21 The number of particles in each cluster of Fig. 5.20 (Reproduced from K. W. Davies & A. E. Hawkins, Proc. 2nd European Symposium on Particle Characterisation, Nürnberg, 1979)

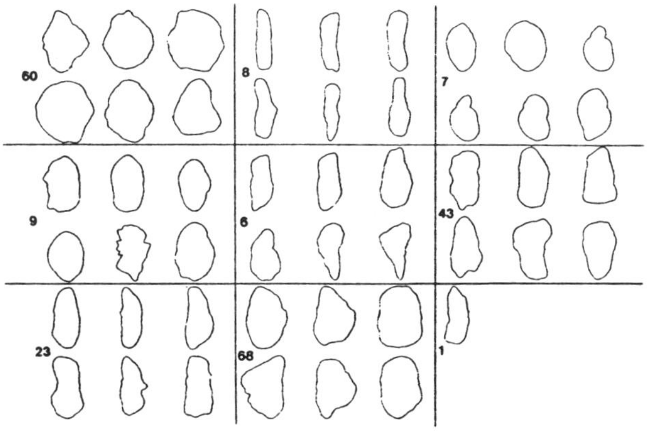

Fig. 5.22 Particle outlines representing nine clusters; numbers show cluster size (Reproduced from K. W. Davies & A. E. Hawkins, Proc. 2nd European Symposium on Particle Characterisation, Nürnberg, 1979)

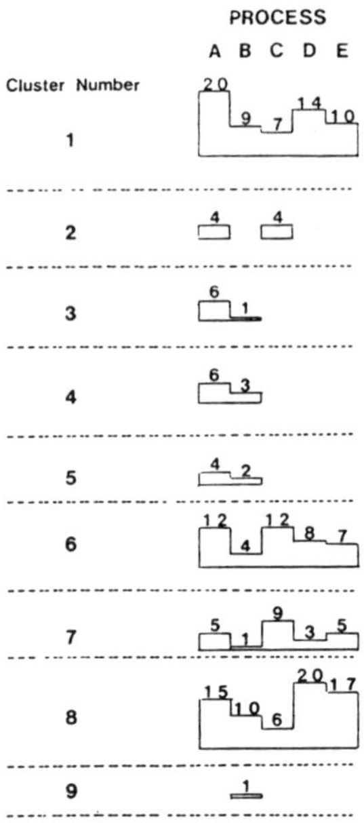

Fig. 5.23 Number of particles from each of five processes in each of nine clusters (Reproduced from K. W. Davies & A. E. Hawkins, Proc. 2nd European Symposium on Particle Characterisation, Nürnberg, 1979)

effectiveness is the percentage of variation explained. This is given by:

$$y = ((T\text{-}W)/T) \times 100\%$$

'It will be near 100% if the clustering is excellent. We can cluster several times and plot y against the number of clusters.'

This can be illustrated by an example given by Hawkins (1980). Some outlines of zircons have been published by Byerly et al., (1975) and these were measured and clustered in the manner just described. Hawkins found that there was a sharp change between five and six clusters, indicating that the return on including more clusters had diminished. This behaviour is illustrated in Fig. 5.24. The paper by Byerly et al. is a valuable one, even though it deals only with the amplitudes of the harmonic series. The authors show how an orthogonal partitioning of chi-square contingency tables revealed that all harmonics except seven and nine carried information and that they contain independent information. They also used a multivariate clustering technique to characterise the shape subpopulations. The only unfortunate

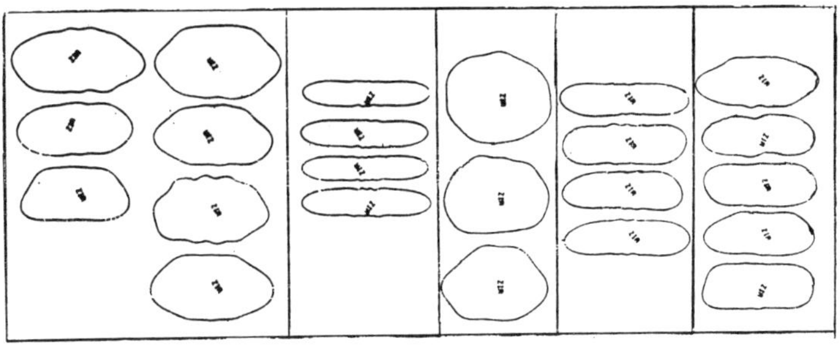

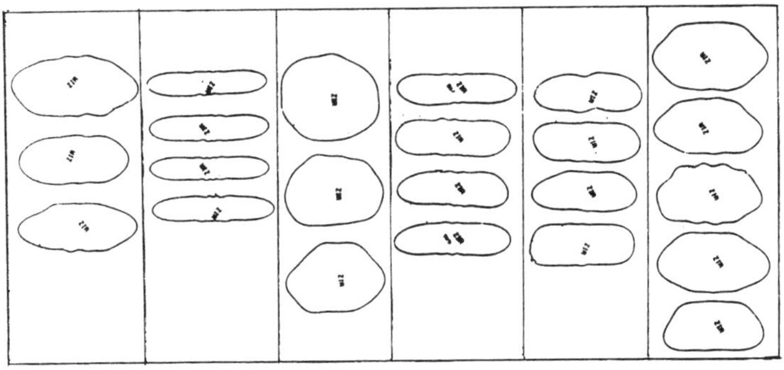

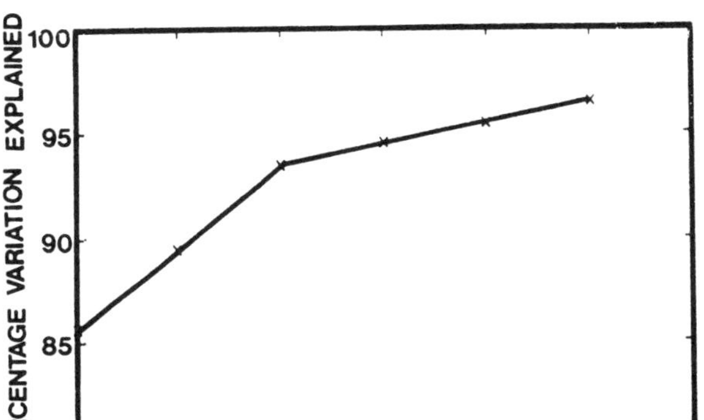

Fig. 5.24 Percentage variation explained and number of clusters used (Reproduced from A. E. Hawkins, Proc. Technical Sessions, Powder Europa, Wiesbaden, 1980)

consideration is the sentence: 'The sum of the harmonics can describe grain shape as precisely as desired.' This is far from correct for obvious reasons, not least the fact that the maximum projections of the zircons were drawn for digitisation.

A text to read, if the clustering technique is to be used, is that by Jain & Dubes (1988). The choice is between hierarchical clustering or partitional clustering; commercial computer packages normally handle the former because of social applications; CLUSTAN was used in the work described above.

Before this Chapter is concluded, mention must be made of the original theoretical work by the latecomers in Meloy's list: Beddow and his group of researchers. Some references have been given already, but a short chronological survey of some key publications should provide guidance into the development of the ideas of this group regarding the use of the terms in harmonic series.

After the initial publication of the independent suggestion that a truncated Fourier-analysis technique could be applied to particle outlines (Beddow & Philip, 1975), the group have tried hard to overcome the disadvantages of the technique. A paper was published by Beddow et al. (1977) that suggested that morphic features could be abstracted in some form from the coefficients in the series. These features were :

$$\text{Centroid Aspect Ratio} = \frac{(\text{maximum diameter through the centroid})}{(\text{dimension perpendicular to the same})}$$

and three terms using the sums of successive groups of harmonic amplitudes:

$$\text{Lumpiness} = \sum_{n=1}^{n_1} A_n$$

$$\text{Roughness} = \sum_{n=n_1}^{n_2} A_n$$

$$\text{Texture} = \sum_{n=n_2}^{n_3} A_n$$

It is interesting to recall that Ehrlich & Weinberg, in their first communication (1970), formulated a similar 'roughness' coefficient.

This work by Beddow et al. may have been a false start, as the pragmatic approach was not mentioned when size and shape terms, of a more fundamental nature were introduced a few years later (Luerkens et al., 1982). Starting from the truncated Fourier series, the authors generate size and shape terms, which, they say, include all the available information about the boundary of a particle profile, with the shape terms both rotationally and size invariant. The size term R_0 is defined :

$$R_0 = (A_0^2 + \tfrac{1}{2}\sum_{n=1}^{\infty} A_n^2)^{\frac{1}{2}}$$

This 'equivalent radius' is the radius of a circle having the same area as that of the particle profile. This value is much more quickly, and generally more accurately, obtained by the direct measurement of outline area in both automatic and semi-automatic image-analysis systems. Davies (1979) has pointed out that a function δ, defined by :

$$\delta = ((\text{outline area}) - \pi(A_0^2 + \tfrac{1}{2}\sum_{n=1}^{m} A_n^2))$$

could be used as a measure of the truncation of the series to m terms. The better the truncation, the closer δ will approach zero.

The shape terms are defined by Luerkens et al. as follows:

$$L_0 = A_0/R_0$$

This is the size-normalised mean radius of the outline with respect to the centroid:

$$L_2(n) = An^2/2R_0^2$$

$$L_3(m,n) = 3(A_m, A_n, A_{m+n} \cos(\phi_{m+n} - \phi_m - \phi_n))/4R_0^3$$

The authors show how to regenerate the particle-outline shape using these terms and also list several possible applications.

This approach was explained again in a wider ranging paper on particle shape (Luerkens et al., 1987). The authors state:

'The raw Fourier coefficients are not useful to describe the profile because they change in value as the profile is rotated. For this reason, they are transformed into invariant morphological and shape descriptors.'

It will be realised that, although the raw data may not be useful to these authors, there is a good deal of literature that indicates that this is not generally true. For example, Holt (1981), when concerned with the outlines of particles produced by crushing, used precisely the same plain plot of harmonic amplitude against harmonic number as had Ehrlich &Weinberg in their first paper. Like other authors, such as Anstey & Delmet referred to above, Holt found useful information in those harmonics that were prominent and in those that were not.

A proposition to consider with regard to the quotation from Luerkens et al. concerns the fact that the phase angles of a harmonic series specifying an outline change as the particle outline is rotated. There is no doubt that they do (although the differences $C_{m,n}$ referred to above do not (Davies, 1979)), and this is given in several papers as the reason for using only the harmonic amplitudes in the research reported. However, the particle whose outline is examined was so situated on the microscope slide; the outline itself would have been different if the particle had rested in a different manner. The point made by Luerkens et al. could equally well be employed to disparage the use of such a statistical diameter as the Feret. The Feret values of the individual particle outlines would be different if the outlines were rotated with respect to the line of measurement.

There is a danger that a straightforward technique may need to be enhanced for parochial reasons: in the case of the Beddow team, partly because of the often very awkward shapes of particles used in powder metallurgy. If this introduces complexity and a heavy demand on computing power, a potential user with a less demanding problem may not utilise what could be useful. However, it is encouraging to find a truncated-Fourier-series treatment being applied to the often difficult outlines of crystals and crystal agglomerates. In what appears to be a new approach, Riebel et al. (1991) use a discrete complex Fourier transformation on 2^8 equidistant support points to define the particle contour. The Fourier coefficients are used in essentially conventional ways, and this may reduce the antipathy to harmonic analysis.

Resolution analysis, the topic of the next Chapter, is another technique in which complexity may have been oversold.

6 Resolution Analysis

This Chapter is not, strictly, about quantifying the shape of a particle outline; it is about quantifying the characteristics of the line making up the outline. Qualitatively, the line could range from very jagged to regular; the surface of the particle might then be described as rough or smooth. For the particle outline, this feature was referred to as 'texture' in the Introduction to this book. A clear example of some of the problems to be faced when dealing with texture is to be found, on the grand scale, in the paper about terrain referred to in Chapter 5 (Stone & Dugundjii, 1965). It will be recalled that these authors say that the roughness of a terrain (or of a curve) is composed of at least three independent factors.

It has been suggested that there is a 'crisis in fine particle characterization posed by rugged boundaries' (Kaye, 1978b). This perhaps overstates the case; but, not unexpectedly, the use of the scanning electron microscope has shown that some particles have more texture on them than can be observed with the optical microscope. Quantitative description of this detail does present problems, but fortunately the need is often a specialised one. In the context of this book, it is as well to keep in mind these following points:

(i) the relatively low number of particles that can be put through an SEM in a day. Although there are methods by which samples can be viewed under the SEM without being covered with a metal conducting film, it is usual to coat with, for example, gold. Although the sputter technique under a vacuum is straightforward, the throughput is limited and adequate sampling will require care.

(ii) the problem of presentation to the direction of observation. This was discussed in Chapter 2. If double-sided adhesive tape is used to mount the particles on the sample stub, the result could be a random presentation; but this might not be so if an epoxy-resin adhesive were used. There is a useful discussion section, after a paper by Whalley & Orford (1982), in which the importance of stage tilting to particle projections is noted. These authors prefer the use of the maximum projection plane.

(iii) the fact that, although great detail may be seen on the electron micrograph, it is the edge that is measured for quantification and this edge may be the sum of overlapping horizons. Of many excellent photomicrographs in the literature, one recent example from a study of adhesion (Lam & Newton, 1991) warrants reference to illustrate the difficulties. The starch, lactose and calcium carbonate particles at around 45μm have very clear visual detail, but, in general, bland profiles will be noted.

The closeness with which the boundary line is to be examined and measured and the detail of the mathematics invoked should be governed by consideration of the quality of the images concerned. In theory, this quality can be very good indeed, particularly with an SEM, because of the depth of field available; but in everyday practice, and particularly with optical microscopy, sharp outlines are not easy to produce. With an SEM, both high magnification and good resolution may be achieved, but using the optical microscope there may be vibration problems. Indifferent focus and, to a lesser degree today, grain of the film are further hazards to good-quality outlines. Further, the image on the film is generally enlarged and printed or back projected.

If direct drawing is used with a microscope, there is the thickness of the drawn line and the control thereof by the operator to be considered, and if direct video analysis is employed there are of course equivalent limits. Close examination of any video screen makes this clear, but the major difficulty is the technical problem of delineating the outline; thresholding is not precise because of limited resolution and shading. This has all become more tractable as video systems have improved, but these matters should be kept in mind when reading much of the published material referred to in this Chapter. As in the history of harmonic analysis there has been a bias in some of the literature towards the mathematical possibilities rather than to the practical realities.

In connection with perimeter measurements, attention was directed, in Chapter 4, to the fact that, as the yardstick size was changed on a large-scale map, the length of a journey measured on the map changed. This, and similar effects of changes of scale of scrutiny, has been known for a long time, but is now given a name: the Steinhaus paradox (Steinhaus, 1954). In his paper Steinhaus writes:

'To measure the length of a river or a highway traced on a map, we employ a transparent sheet with a family of equidistant parallels L_i (i = ..., -2,-1,0,1,2...). The arc A to be measured cuts L_i in a_i points, and $S_o = \Sigma a_i$ is the number of intersections. Turning the sheet through an angle $\pi k/m$ (k = 0,1,....,m-1) we get s_k intersections and the grand total is:

$$N = \sum_{k=1}^{m-1} s_k$$

Calling d the distance $L_i\ L_i+_1$, we get the expression $Nd\pi/2m$ as an approximation of the length of A.'

Steinhaus notes later:

'It is much easier to count the intersections of the straight lines of the transparent sheet than to drive a measuring wheel along the sinuosities of the curves or to pace along them with a compass; the results obtained by these familiar methods are more liable to personal and instrumental errors than those obtained in the same time by our method.'

In the course of his paper, Steinhaus refers to quantitative comparison of the frontier of Poland given in a map in the year 963 with that given in the maps of today. The problem is that there are differences of detail provided. 'The same difficulty arises when measuring such objects as contours of leaves or perimeters of plane sections of trees: the result depends appreciably on the precision of the instruments employed.'

The subject of frontiers arises in a book on the causes of war published a few years after the Steinhaus paper. This was written by I.F.Richardson but published posthumously (1960) and was followed shortly by an appendix to it (Richardson, 1961). This latter 60 pages contained a detailed mathematical analysis of the techniques of mapping populations, of topological theorems and of the problems of contiguity; it was mentioned in Chapter 4 in connection with 'compactness'. When dealing with networks of cells, Richardson took integrals around simple geometrical figures as a preliminary to taking them around frontiers shown on political maps. An embarrassing doubt arose as to whether actual frontiers were so intricate as to invalidate that otherwise promising theory. A special investigation was made to settle this question.' In two pages describing this investigation, Richardson triggered, through an intermediary, a great deal of research in many fields of science. If he had made use of the work of Steinhaus, it might not have happened.

In his appendix, Richardson shows that, when coastlines or borders between countries are measured using smaller and smaller units of measurement, then, when logarithms of the apparent length (L) and the step length (s) are plotted jointly, the relationship between log (length) and log (step length) is apparently linear. Moreover, the more ragged the coastline or border, the steeper is the log-log plot. Richardson's illustration is reproduced in Fig. 6.1. Richardson summarises his findings 'by the useful empirical formula $\Sigma l \propto l^{-\alpha}$ where Σl is the total polygonal length, l is the length of the side of the polygon, and α is a positive constant, characteristic of the frontier.'

This is as far as Richardson takes the topic, but there appears to be some confusion about the work. Mandelbrot (1967) and some later authors refer wrongly to what Richardson actually wrote, and Orford & Whalley (1983) refer to 'Richardson's unpublished study from the beginning of the twentieth century.' For many irregular lines, Richardson's empirical finding seems to be true. It may be exhibited in two ways: the original Richardson plot of log L versus log s, or the

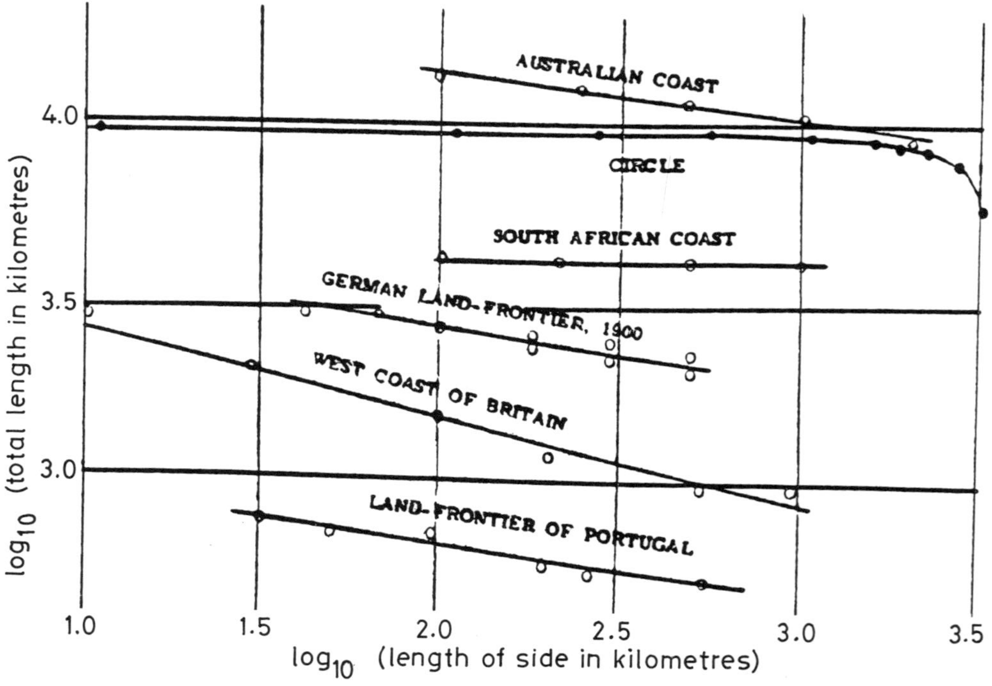

Fig. 6.1 Double logarithmic plot of apparent length versus step length for frontiers and coasts
(Reproduced from L. F. Richardson, 'General Systems Yearbook', 1961, **6**, pp.139-187, by permission of the International Society for the Systems Sciences.

alternative log n versus log s, where n is the number of steps of size s. If linear plots appear to be obtained, then, in the first form, the equation:

$$\log L = \log A - b \log s$$

should apply. For reasons that will appear later, this equation is generally written as:

$$\log L = \log A + (1-D) \log s \qquad \text{...(6.1)}$$

In the second form, the equation:

$$\log n = \log B - D \log s \qquad \text{...(6.2)}$$

should apply. If these equations apply, the consequence is that there is a power-law relation between L and s or n and s. This is important and neglected in much of the powder literature.

In a thoughtful and informative paper, Farin et al. (1985a) describe the way in which the estimated length grows with a decrease in yardstick size (Fig. 6.2). They say: ‘The rate of appearance of new features of the irregularity as a function of the size of the probing yardstick (or degree of magnification) is measured.’ The authors give the power law $n \propto r^{-D}$ where n is the number of yardsticks of size r needed to measure the total length of the wiggly line.

The use of discs rather than chords is one of several variations in resolution analysis; another is the use of squares, as illustrated in Fig. 6.3. taken from the book by Feder (1988). Feder states that the number of such squares of side δ needed to cover the coastline on the map is roughly equal to the number of steps used when walking dividers with an opening δ along the coast. Fig. 6.4 shows how Feder found that the measured length increased as the yardstick was reduced. The slope is far steeper than even that for the west coast of Britain.

The significance of the Richardson paper in the context of this book is mainly that, if such a power law applies to a wiggly line such as that in Fig. 6.5(a), essentially the same power law will apply if the line is bent round to form the closed loop of Fig. 6.5(b). In principle, there is here a measure for the texture of a particle outline. Notice, however, that the measure is only of the line. The same wiggly line with approximately the same numerical value could be bent round to produce the outline of the elongated form of Fig. 6.5(c) or into a triangular form or a dumb-bell form. This is not brought out in some titles of papers nor, indeed, in the papers. This may not matter if the application is to particle growth, for example, but it could be important in many other applications including powder flow and packing.

In addition to this restriction to texture, it will be noted that it is a single-number measure. From what has been the case in earlier systems with a single-number measure, it is to be expected that this resolution analysis could also lead to ambiguity. This is so, and can be illustrated by the simple example of the line in Fig. 6.6(a).

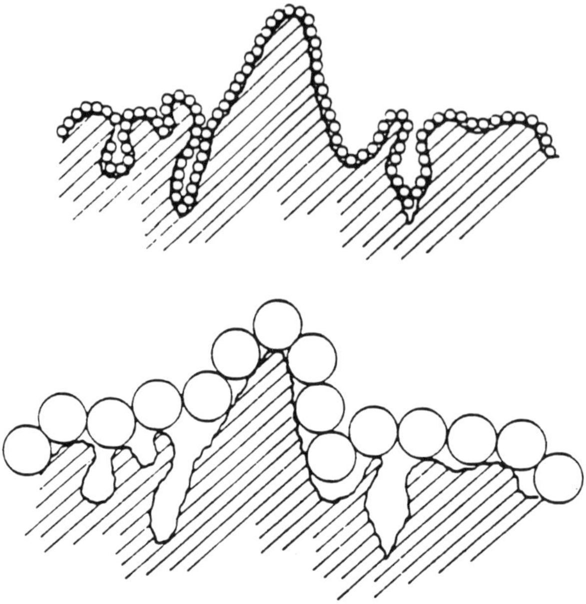

Fig. 6.2 An illustration of the manner in which the detail obtained of a convoluted line increases with decrease of the yardstick size
(Reproduced from D. Farin et al., *Langmuir,* 1985a, **1,** pp.399-407, Copyright 1985, American Chemical Society, reprinted with permission)

So far as the measurement of the apparent length is concerned, the line could be divided up differently, as in Fig. 6.6(b) or in Figure 6.6(c). Some of the published material can also be used to show ambiguity. For example, there are outlines given by Farin et al. (1985) and by Kaye (1989) for which the log-log plots have similar slopes, but where the line structures are visibly different. This ambiguity means, of course, that the eqns (6.1) and (6.2) for a line texture cannot be used to redraw that line texture, and must be used with care to compare members of 'families' and not of widely varying populations. It would seem to be particularly useful for the study of growth or erosion. For example, Brady & Ball (1984) found the principles valuable in a study of the electrolytic deposition of copper in diffusion-limited conditions.

The significance of the Richardson paper referred to earlier was not appreciated until a short note was published calling attention to it (Mandelbrot,1967). Not much attention appears to have been paid to this note either, nor does it feature in some lists of references, but the three books that followed it (Mandelbrot, 1975, 1977, 1982) have had far reaching effects in mathematics and many other branches of science. It is in these books that Mandelbrot explains in detail the concept of

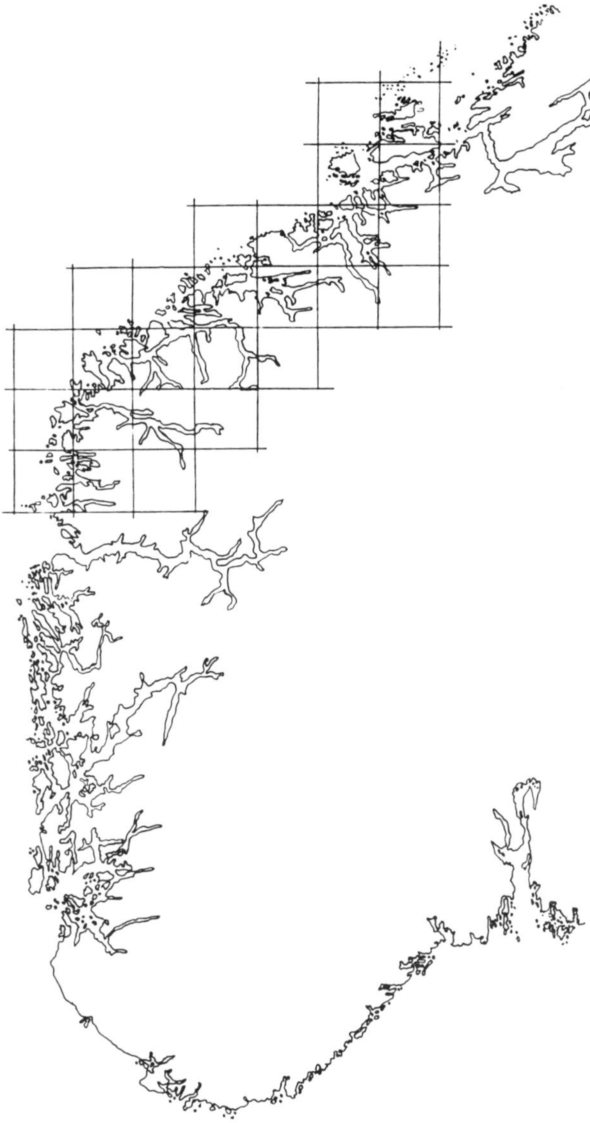

Fig. 6.3 The use of a square grid to obtain the data for resolution analysis (Reproduced from J. Feder, 'Fractals', by permission of the author and Plenum Press, N.Y., 1988)

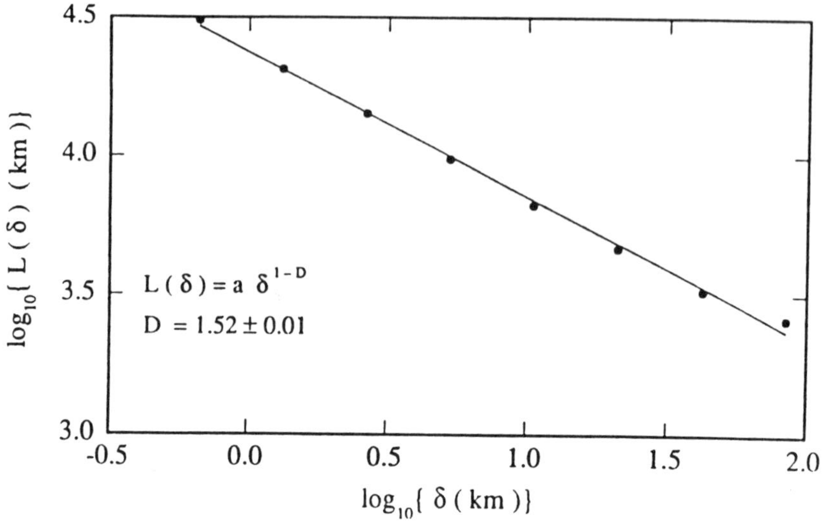

Fig. 6.4 The double-logarithmic plot resulting from the measurements in Fig. 6.3 (Reproduced from J. Feder, 'Fractals', by permission of the author and Plenum Press, N.Y., 1988)

non-Euclidean fractal geometry and of fractal dimensions in particular. This is the geometry of lines and surfaces that used to be considered by mathematicians as 'pathological' because continuous functions could not be fitted to them.

According to Falconer (1990), 'the word 'fractal' was coined by Mandelbrot in his fundamental essay from the Latin *fractus,* meaning broken, to describe objects that were too irregular to fit into a traditional setting.' Falconer continues 'any fractal worthy of the name will have a fine structure, i.e. detail at all scales. Many fractals have some degree of self-similarity – they are made up of parts that resemble the whole in some way. Sometimes, the resemblance may be weaker than strict geometrical similarity; for example, the similarity may be approximate or statistical.'

The constant D in eqns. (6.1) and (6.2) is the 'fractal dimension', and has been called a renaming of the Hausdorff dimension by Rys & Waldvogel (1986). These authors state:

'The fractal dimension D characterises any self-similar system: Upon changing the linear dimension by a scale factor f, a fractal quantity (such as a contour or a surface) changes by the factor f^D for any value of f.'

The work by Hausdorff is dealt with thoroughly in the first part of the text by Falconer.

(a)

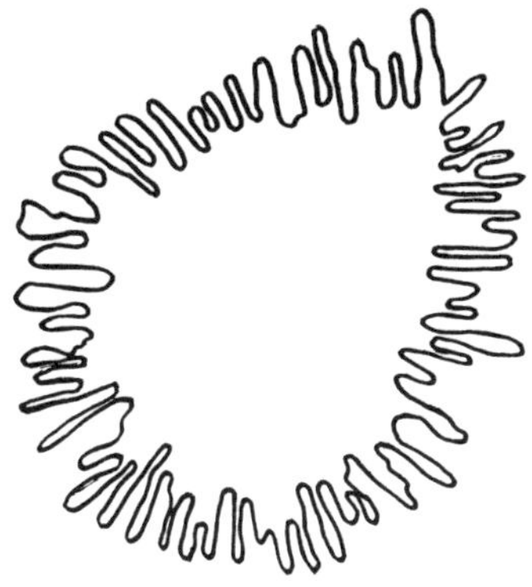

(b)

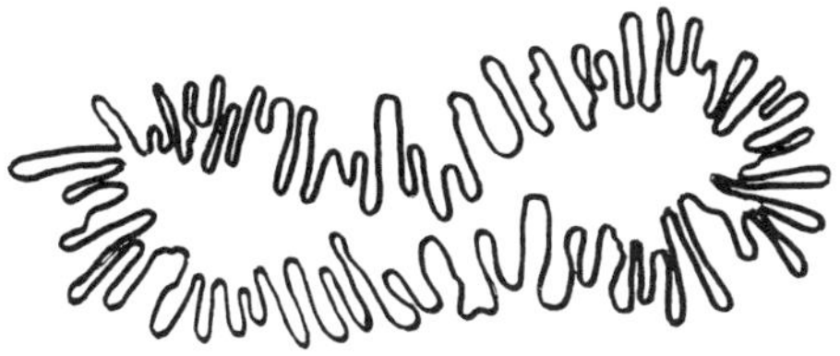

(c)

Fig. 6.5 A convoluted line in three arrangements

In a paper on landscapes, which particle outlines are not unlike even though on a small scale, Burrough (1981) summarises the subject succinctly :

'Mandelbrot (1977) has introduced the term 'fractal' specifically for temporal or spatial phenomena that are continuous but not differentiable, and that exhibit correlations over many scales. The term fractal strictly defined refers to a series in which the Hausdorf-Besicovitch dimension exceeds the topological dimension.

A continuous series, such as a polynomial, is differentiable because it can be split up into an infinite number of absolutely smooth straight lines. A non-differentiable continuous series cannot be so resolved. Every attempt to split it up into smaller parts results in the resolution of still more structure or roughness. For a linear fractal function the Hausdorf-Besicovitch dimension D may vary between 1 (completely differentiable) and 2 (so rough and irregular that it effectively takes up the whole of a two-dimensional topological space).'

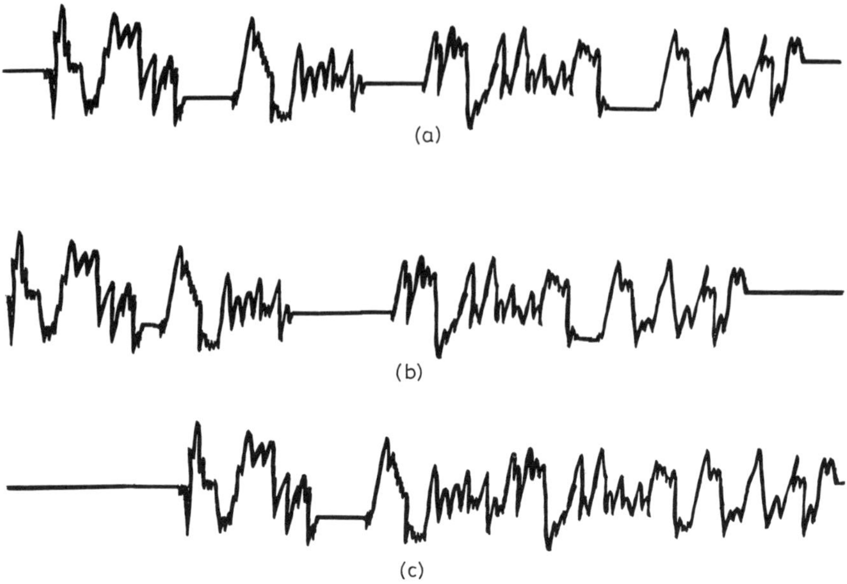

Fig. 6.6 A convoluted line divided up in three different ways

A relatively early paper (Berry & Lewis, 1980) on the Weierstrass-Mandelbrot fractal function states that, although it is continuous everywhere, the function is differentiable nowhere. The fractal nature implies that repeated magnification of its graph reveals ever-finer levels of detail. Such behaviour, the authors continue, contrasts sharply with that of a differentiable function, whose graph tends to a straight line when magnified.

In his recent text Falconer states:

'A glance at the recent physics literature shows the variety of natural objects that are described as fractals – cloud boundaries, topographical surfaces, coastlines, turbulence in fluids, and so on. None of these are actual fractals, their fractal features disappear if they are viewed at sufficiently small scales. Nevertheless, over certain ranges of scale they appear very much like fractals, and at such scales may usefully be regarded as such. The distinction between 'natural fractals' and the mathematical 'fractal sets' that might be used to describe them was emphasized in Mandelbrot's original essay, but this distinction seems to have become somewhat blurred. There are no true fractals in nature (there are no true straight lines or circles either!)'

According to Feder (1988), the definition for a fractal proposed by Mandelbrot is now 'a fractal is a shape made of parts similar to the whole in some way'. Feder considers that this definition 'contains the essential feature: a fractal looks the same whatever the scale. Look at some nice cumulus clouds, for example. They consist of big heaps with smaller bulges that have smaller bumps with bumps on them and so on down to the smallest scale you can resolve. In fact, from a picture showing only the clouds one cannot estimate the size of the clouds without extra information.'

It should be noted that Feder provides a good reference list, including many papers in which Mandelbrot is a coauthor.

Self-similarity is the condition where the image of the line boundary, for example, or surface appears the same under increasing magnification; the 'pattern' is independent of magnification. This means that a photograph cannot be identified as being from magnification *m*, 10*m*, 100*m* or any other. This is illustrated in several texts; a photograph of a minaret cauliflower (Kjems, 1991) is a classic example, but it is certainly nowhere near true for many particle outlines.

The first powder scientist to realise that Mandelbrot's ideas about Richardson's findings might be applied to the outlines of particles was Kaye, who presented his initial ideas at a meeting in 1977 (Kaye, 1978a) and in a more substantial form in a paper (Kaye, 1978b). Richardson had walked dividers along his maps to obtain his estimates of length using different spacings of the legs. Kaye used the same technique on real and made-up particle outlines and obtained reasonable straight lines on log-log plots. Since these first communications, Kaye has published several papers and chapters of books on the topic (Kaye, 1981, 1984, 1986; Kaye & Clarke, 1986; Kaye et al. 1986). It is interesting that, in the first paper, Kaye speculates on measuring line length by a method of the Steinhaus type. The most straightforward access to this published work and to Kaye's ideas on the development and application of the subject is through his book (Kaye, 1989).

Kaye transforms the perimeter estimates and stride lengths into dimensionless form by normalising with the maximum Feret's diameter before plotting his double-log graphs. Kaye refers to these plots as 'Richardson plots'; others have referred to them as 'Mandelbrot-Richardson plots' (Schwarz & Exner, 1980).

The log-log plot is interesting on two counts. On the one hand, such a plot is notorious for producing apparently straight lines and yet the plot is not linear for some of the outlines measured and published in the literature. Secondly, there is scatter about nearly all the published lines. The authors of most of these publications suggest that this scatter is due to experimental error, and so use least-squares linear regression to fit the, so-called, fractal line to the data. This is a natural procedure to adopt, but is difficult to justify on grounds other than expediency. A straight line obtained on a log-log plot is not of importance in itself; as was pointed out earlier, the straight line indicates the possibility that there is a power relation between L and s or between n and s. It is this that is important, and the power index and the intercept should be optimised from the original data. This can be done these days very precisely now that computers facilitate the use of numerical analysis (Bunday,1984).

It has been written (Mark & Church,1977) that:

'A common problem in empirical science is to select a precise curve to describe the covariation of two variables from a set of paired samples. There can be a number of reasons for wishing to fit such a curve: (1) to provide a basis for predicting values of one variable, given values of the other; (2) to calibrate a new instrument in terms of an established one; or (3) to determine the parameters of the curve for descriptive purposes or for comparison with theory or other empirical results. Least-squares regression is used almost universally to achieve this. However, it has been shown that regression analysis should be used only for purposes of prediction (see Lindley, 1947; that author reviews the long history of the problem). The related statistical technique termed *functional analysis* should be used to provide unbiased estimates of the parameters required in situations (2) and (3).'

The authors give some examples, provide references and outline the mathematical treatment required, but it is not appropriate to reproduce that here. It is worth mentioning, however, that, in the discussion of a paper using Richardson plots (Whalley & Orford, 1982), the authors state that some bias must occur in the calculation of the slope of the log-log plot when a weighted least-squares procedure is not used.

The unquestioned use of the 'Richardson plot' seems to be echoed in the unquestioned invocation of fractal mathematics into the resolution analysis of the texture of particle outlines. The outline of a particle does not exhibit self-similarity, i.e. there is no exact mathematical similarity at all magnifications. Kaye (1989) removes the problem by invoking 'statistical self-similarity': 'Statistical self-similarity means that although the magnified structure of any local part of the boundary is not exactly self-similar, any one section of the boundary looks like any other 'snippet' taken from different parts of the profile.' This is not how Mandelbrot defined statistical self-similarity (Mandelbrot,1967): 'Highly involved curves such that each of their portions can – in a statistical sense – be considered a reduced-scale image of the whole.'

Kaye illustrates his meaning of statistical self-similarity with a series of profiles originally published to show certain aspects of harmonic analysis. Like many of the profiles given in the powder literature to which fractal geometry has been applied, these outlines are clearly differentiable: a tangent can be drawn to the curve at any point.

The fact that natural boundary lines are not truly fractal was dealt with in the paper by Farin et al. (1985a) already mentioned. In this paper, the authors give these following conclusions:

'(a) A boundary line that obeys the equation $n \propto r^{-D}$ need not be self-similar. (b) Non-statistical strict self-similarity is scarce. (c) Even in the absence of self-similarity, the exponent D in the equation $n \propto r^{-D}$ carries useful information in the original spirit of Richardson coast-line analyses, i.e. an empirical relation between scales and measured object. However, (d) The equation $n \alpha r^{-D}$ is quite insensitive for detailed analysis of local short-range resolution variations. It must be replaced by more sensitive methods. (e) The equation $n \alpha r^{-D}$ with $D \langle 1.2$ describes well virtually all low-irregularity lines tested by us and others, even lines that are clearly not self-similar.'

This paper is not well cited in the powder literature but may be considered essential reading for anyone intending to use resolution analysis, and was probably the first to use the expression.

The lack of sensitivity is illustrated by Farin et al. using eqn. (6.2), but is equally obvious using eqn. (6.1). In Fig. 6.7, log L versus log s for D values of 1.11, 1.13, and 1.15 is plotted. This lack of sensitivity would seem to support the need for numerical analysis mentioned earlier; but Barnsley, for example, in his book on fractals (1988) states that 'regardless of how much experimental data we have, and regardless of how many scales of observation are available to us, we will always end up estimating the slope of a straight line corresponding to a finite range of scales.'

One potentially important use of this type of resolution analysis is the application to some forms of fracture. For the powder scientist this could refer to the breaking of tablets, but it is of wider application. It returns to the original measurements of coastlines with the linear log-log plot.

Chermant & Coster in a review of fractography (1979) discuss the problems connected with the quantitative aspects of the morphology of a fracture line. After explaining earlier attempts to quantify the waviness, the authors suggest that the fractal dimension of the line of fracture be calculated. This is a follow-up to the reports of earlier exploratory experiments (Chermant & Coster, 1978; Coster & Deschanvres, 1978).

Chermant & Coster write:

'For a fracture line, the fractal dimension lies between 1 and 2. The value is 1 for regular curves, with a defined tangent at all points. The value tends to 2 when the line fills all the space.' They go on to say that 'even if the fractal dimension of a fracture cannot be defined, because too few phenomena are responsible for its irregularity, it is still interesting to have the double logarithmic plot, because the slope, at a given point, is a function of the irregularity of the fracture line for the given scale.'

The authors illustrate the effect of the line of grain boundaries followed by the fracture on the log-log plot in Fig. 6.8.

The double-logarithmic plot is not truly linear.

Flook (1978, 1979) and, later, Kaye have published papers that use the double-logarithmic plot to characterise what appear to be two separate features of some particle outlines. In his first paper, Flook characterised silhouettes from the Medalia paper (1970/71), discussed in Chapter 4. Both Flook and Kaye found that, with

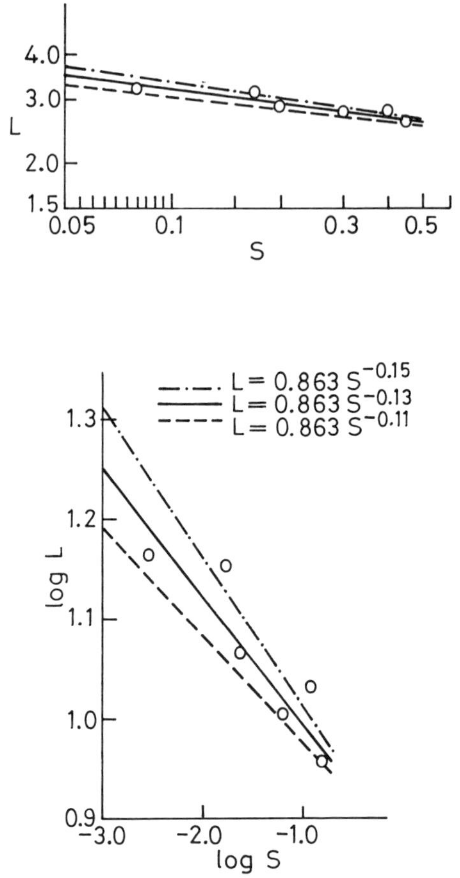

Fig. 6.7 Double-logarithmic plots to illustrate:

(i) the flattering impression of linearity given by a cramped scale when printed in the literature

(ii) the lack of sensitivity of the plotting technique

some outlines, a more steeply sloping line was obtained with the larger measuring sticks and a less steeply sloping line was obtained at a close level of inspection. These were referred to as the 'structural fractal line' and the 'textural fractal line' respectively. This detail was not observed in the original pioneer work by Kaye (1977), using the manual technique with dividers. This was a consequence of using the least-squares technique with few points on the log-log graph. Flook used an automatic image analyser, as indeed did Chermant & Coster (1978) in a similar

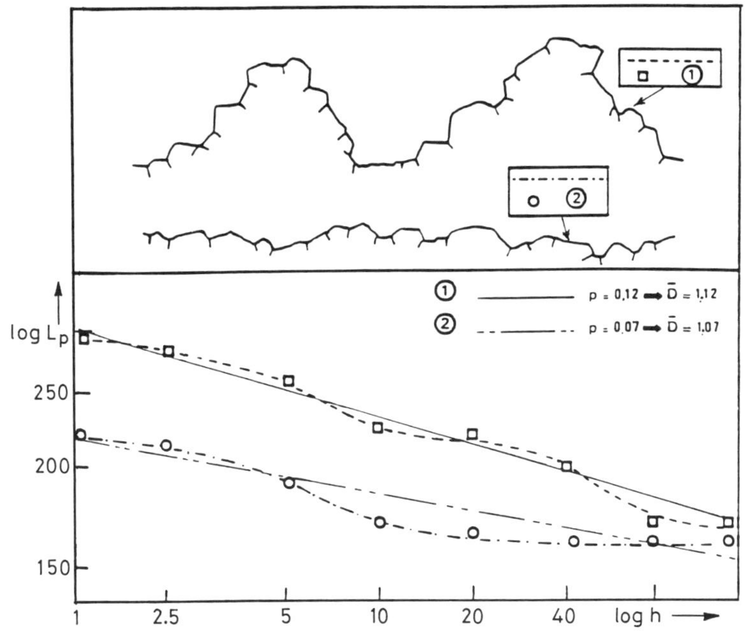

Fig.6.8 The effect of grain boundaries on a double-logarithmic plot of apparent length versus step length
(Reproduced from J. L. Chermant & M. Coster, *J. Materi Sci.*, 1979, **14,** pp.509-534, by permission of Chapman & Hall, London)

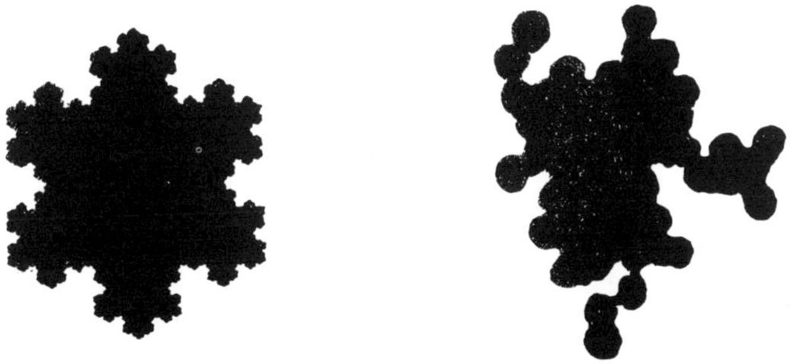

Fig. 6.9 Test figures used to compare three methods of measurement:
(a) = Kock, (b) = Flook
(Reproduced from H. Schwarz & H. E. Exner, *Powder Technol.*, 1980, **27,** pp.207-213, by permission of the authors and Elsevier Sequoia S.A.)

manner for their contemporaneous studies on some of the outlines published by Hausner (1966). Although brief, the introduction to the paper by Chermant & Coster is particularly lucid about the mathematical background of the work.

There is an interesting paper (Schwarz & Exner, 1980) in which three outlines are measured using three different methods and the fractal dimensions calculated from the resulting double-logarithmic plots by straight lines of best fit using a linear-regression analysis. The test figures are reproduced in Figs. 6.9 and 6.10. The values of fractal dimension (calculated over a restricted range of resolutions) are given in Table 6.1.

Table 6.1 –Values of fractal dimensions for test figures as determined by different algorithms. The numbers in brackets give the range of resolution as a fraction of the maximum particle dimension.

Test figure	Method 1 (divider)	Method 2 (Flook)	Method 3 (Videoplan)
1 (Medalia)	1.09 (0.023-0.23)	1.37 (0.11-0.2)	1.20 (0.085-0.25)
2 (Koch)	1.25 (0.029-0.22)	1.26 (0.02-0.25)	1.22 (0.085-0.25)
3 (Flook)	1.24 (0.024-0.24)	1.32 (0.055-0.18)	1.35 (0.085-0.25)

(Reproduced from H. Schwarz & H. E. Exner (1980) by permission of the authors and Elsevier Sequoia S.A.)

For the 'Koch Triadic Island', for which the fractal dimension is known from theory to be 1.2618, the three experimental methods give values reasonably closely in agreement; for the other two outlines, the agreement is not good. Method 1 was a semiautomatic version of the walking-dividers system evaluated manually by Schwarz & Exner on their particular semiautomatic analyser using a digitising tablet. This was time-consuming and exhausting for the operator. Method 2 was described by Flook in his paper (1978): 'In this method a curve is considered to be made from a series of closely spaced points. A series of overlapping circles of radius η is drawn with their centres on each of the points of the curve in turn, thus describing a path of width 2η covering the curve. The area of this path divided by its width gives an estimate of the length of the curve. As η increases, the circles have a greater degree of overlap and obscure more and more of the fine details of the curve, and the length estimate $L(\eta)$ will decrease. The fractal dimension D may be evaluated from the doubly logarithmic plot of $L(\eta)$ *versus* η'

Method 3 was introduced by Schwarz & Exner using the digitising tablet, but with computer assistance. An initial string of profile coordinates at constant spacing was obtained with the digitiser following the profile with the instrument cursor. In the calculation, the profile length is calculated for all the coordinate points, then for every second pair of coordinates, then for every third pair, and so on. To obtain a mean measuring stick length for each estimate, the profile-length estimate is divided by the number of distances counted. (Schwarz & Exner used eqn. (6.1) rather than eqn. (6.2). The use of the latter would have eliminated this last calculation step.)

It is clear that different methods of determining fractal dimensions may give different values, and care must be taken in reaching conclusions. For example, Farin et al. (1985a) question the validity of the measurement of distances along the analysed line used by Schwarz & Exner. Farin et al. used automatic digital processing to obtain data equivalent to those from the original Richardson technique, but plot their results differently,as indicated in Fig. 6.10(a) (D = 1.10) for the Medalia carbon-black aggregate.

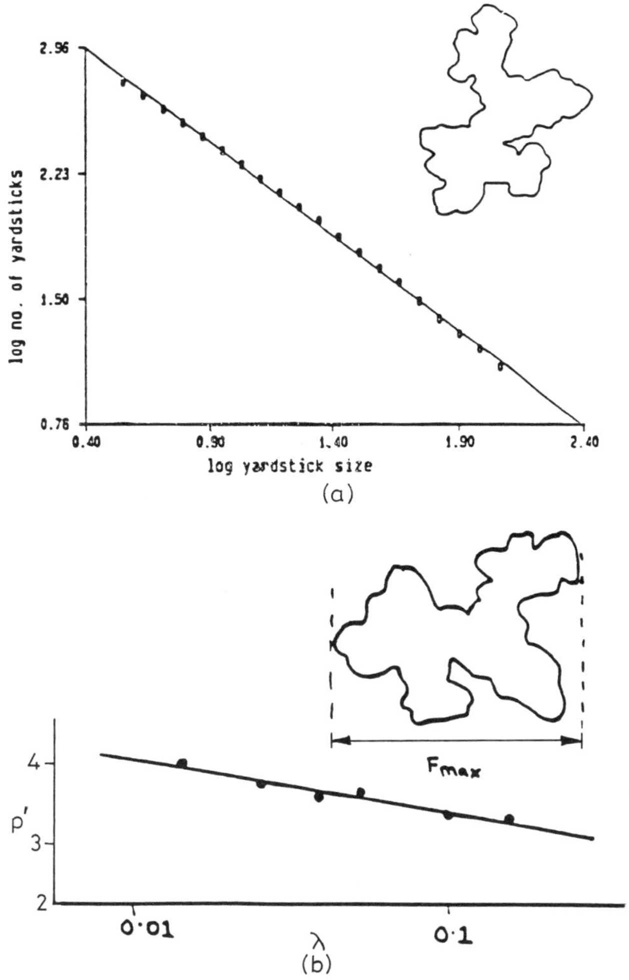

Fig. 6.10 Two different measurements of a carbon-black aggregate
(Reproduced from (a) D. Farin et al., Langmuir, 1985a, **1,** pp399-407, Copyright 1985, American Chemical Society, reprinted with permission; (b) B. H. Kaye, Powder Technol., 1978b, **21,** pp.1-16, by permission of the author and Elsevier Sequoia S.A.)

This may be compared with the results for the same outline published by Schwarz & Exner using a full range of resolutions *(D* = 1.13), by Flook (*D* = 1.07 and 1.37), and by Kaye in Fig. 6.10(b) (*D* = 1.19).

Another useful paper on methods for obtaining the data with which to apply resolution analysis in automatic image analysis is that by Clark (1986a). Three techniques are described, the most important one approximating to the hand-and-dividers method, but avoiding the varying step length of the Schwarz & Exner algorithm. This work has been followed up by a further method that is claimed to be quicker (Kennedy & Lui, 1986).

Avnir et al. (1985), in one of their several papers on surface-geometric irregularity, treated quantitatively by fractal geometry, state:

'If a surface is experimentally found to be fractal, then one has an important clue to the mechanism by which that object was formed: over a certain range of scales, an iterative process, or a fractal process operated. We regard this consequence as one of the future key applications to our fractal approach.'

For this and similar reasons, the use of both resolution and fractal analysis has developed in many disciplines; because of their potential strengths, misuse could seriously mislead the user. Kaye, in his book (1989), makes a particular effort to warn the reader of this danger as well as illustrating uses of the 'Richardson plot'.

In addition to the points made by Kaye, it is prudent to keep in mind that the dimension D obtained for a wiggly line does not permit a reconstruction of that line to be made. The mathematics permits only the construction of a line that *looks like* the original. Batty, in a set of three articles indicating the use of fractals in geography (1992), employs the exact word needed;

'Starting from a two-dimensional base, it is possible to generate highly realistic models of landscapes. Such scenes are, of course, forgeries.'

For modelling the behaviour of particles, alone or in bulk, "forgeries" of outlines may be valuable, but they are not the actual outline shapes that were measured.

In Chapter 3 it was suggested that experiments with shape-sorted material might be better suited to the readers' needs than shape measurement. The work reported by Avnir and his coworkers around the world is another example of the use of bulk material rather than the individual particle. The paper by Avnir et al., just referred to, reports the determination of surface geometric irregularities of a wide variety of particulate materials. This was achieved by the reanalysis and reinterpretation of both previously published and newly obtained experimental adsorption data using the fractal theory of surface science. In brief, a powder would be sieved into several fractions and for each fraction the apparent monolayer value determined by a standard method. The change of apparent monolayer value n with change in average radius of a spheroidal particle is given, perhaps, by $n \propto R^{(D-3)}$ where D is the fractal dimension, and it was found to be so for quite a wide range of materials; but the authors state that for this equation to hold, the particles should be similar to one another, and they go on to explain the concept.

For this book, it is more interesting to illustrate 'particle similarity' by a quotation from a paper about coal particles published nearly 60 years ago (Needham & Hill, 1935):

'Another observation of considerable importance relates to the 'outline' or silhouette of the particles. It was found that in all fractions the same assortment of shapes, so far as could be judged from projections, occurred.

When particles are examined by projection through a microscope it is impossible to tell from the appearance of the silhouettes what size fraction is being observed, since, for example, 150 to 170 mesh (Tyler) particles, magnified 500 diameters, are indistinguishable from 32 to 35 mesh magnified 100 diameters. Moreover, this similarity in appearance of projection extends to the larger fractions up to 1½ in. The relative thickness of the particles, which is not observed in projections, decreases with decrease in size. Thus 6 to 7 mesh (about ⅛ in.) particles present the same assortment of outlines as 1½ in. to 1 in. pieces, but are found to be relatively thinner when closely examined.'

It is possible that some materials in powder science will exhibit self-similarity, for example when the material has been through a mill to produce a powder, and the use of the term 'fractal' may then be justified. However, many outlines will not exhibit self-similarity, but the application of resolution analysis will still be possible; in these cases, the term 'fractal' would seem inappropriate.

An example of an unsuitable use of 'fractal' is to be found in a stimulating paper (Clark, 1987) that introduces a method to overcome the disadvantage of resolution analysis mentioned earlier: i.e. the fact that 'it provides little information on the macroscopic particle shape'. Clark extends the original hand-and-dividers method for resolution analysis. Instead of terminating the polygon with a generally unequal closing side, the silhouette edge is continuously traversed. This is illustrated in Fig. 6.11. According to Clark, 'for all but the most recalcitrant silhouettes at certain step lengths, the dividers very soon will start to walk in their own footsteps, the points falling on the vertices of a regular polygon'. The regular polygons have been discussed elsewhere (Clark, 1986b). If a different step length is used, a different polygon will result, and the author names the series of polygons that result from a series of step lengths: 'polygonal harmonics' and 'fractal harmonics'. The first harmonic to arise, with step lengths of about the particle diameter, consists of two diametrically opposed points. The next is a triangle (third harmonic), and then a rhombus (fourth), a pentagon, and higher-order harmonics. Clark suggests that the scheme be used 'primarily to deal with rugged particles which defy simple geometric description'; to describe the macroscopic shape by fractal harmonics; and to describe the microscopic shape by fractal analysis.

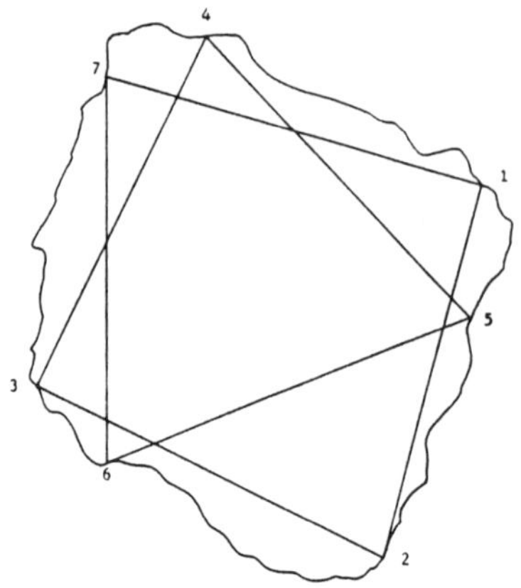

Fig. 6.11 A step length that required two traverses of the outline to achieve closure (Reproduced from N. N. Clarke, *Powder Technol.*, 1987, **51,** pp.243-249, by permission of the author and Elsevier Sequoia S.A.)

The idea has been investigated using computer-generated and real particle cross-sections (Young et al., 1990). These authors used the expression 'polygonal harmonics' because the technique is not restricted to particles displaying fractal geometry. They suggest that the technique is ideally suited to very jagged and highly re-entrant particle cross-sections that are difficult to analyse using conventional Fourier analysis; what is more, fractal analysis measures only surface roughness, whereas polygonal harmonic analysis deals with overall outline shape'. This is true, but, if the paper is studied, it will be realised that the overall outline shape is indicated only by the numbers generated.

In two recent papers (Reilly & Clark, 1991; Maeder & Clark, 1991), the potential of polygonal harmonics is developed and explained. Also, in the paper by Reilly & Clark, a 'surface roughness index' is introduced that is given by (1-s), where s is the negative slope of a Richardson plot. The authors write:

'The surface roughness index described below is found in the same way as fractal dimension using the divider stepping technique. However, the authors have refrained from using the term 'fractal dimension' only because many of the objects are euclidean rather than fractal in nature. Those not constrained by this rigour may regard this roughness index as fractal dimension.'

It would seem more sensible to use the slope s as the roughness index, in the same way as Chermant & Coster suggested in their review paper of 1979 mentioned earlier and as Richardson implied in his original communication.

An alternative to polygonal harmonics, the use of harmonic analysis for the description of the macroscopic shape, is said by Clark (1986a) to be 'unsuitable for dealing with jagged surfaces, which are not easily described by periodic sine waves.' This view may be compared with the results given in a paper dealing with scanning-electron-microscopy images (Whalley & Orford, 1982). Resolution analysis was applied to very angular images. These authors note: 'Fractals are best suited to the highly indented outlines, as they are not very sensitive to small changes in outline variation. Conversely, Fourier methods are appropriate to characterize small changes in outline.' They suggest: 'Fractal and Fourier techniques are complementary in establishing particle form and roughness', but note elsewhere in the paper: 'the problem of fractal analysis is that a single valued function is always limited in the amount of information it can convey about the form which is being approximated'.

The following year, these authors published a similar, but more detailed, paper on the use of the fractal dimension (Orford & Whalley, 1983). They concluded:

'Given the value of harmonic analysis with respect to shape studies on relatively non-angular particles, and given the success of fractal analysis in characterizing very irregular particles, then the two techniques can be viewed as complementary, such that an adequate specification of particle form is now possible for all points on the continuum of particle form.'

Fractals are both intellectually fascinating and very important in areas outside particle science and technology. The range is from, in commerce, fractal images appearing on T-shirts *(Physics World,* 1991) and in films (Ivry, 1990), to physics with, for example, the analysis of the fractal shape of severe convective clouds (Rys & Waldvogel, 1986), an earlier paper on rain and cloud areas (Lovejoy, 1982), and a model for dense packings of discs rolling on one another that might have application for turbulence and tectonic motion (Herrman et al., 1990). In physical chemistry, the application of fractal models to describe luminescence quenching in water-in-oil microemulsions has been extended to electric percolation (Modes et al., 1990) and, in geology, fractal mathematics is put forward for application to the suggestion that measured sedimentation rates are slow when measured over long time spans and rapid when measured over short time spans (Plotnick, 1986).

A recent collection of ten articles (Bunde & Havlin, 1991) has within it detailed accounts of fractal mathematics applied to percolation, growth and fracture among other topics. There are included both self-similar fractals, which are invariant under isotropic enlargement, and self-affine fractals, which are statistically invariant under anisotropic enlargement.

Another interesting paper is that by Thiébaut (1989). This author introduces fractal geometry to the field of computer-performance prediction and uses the fractal dimension as a simple means of measuring program behaviour. In the context of this Chapter, this quotation is apposite:

'Scale-invariance applies in an appealing fashion to the study of program performance. If a program is modeled by a fractal random walk, the clusters that it creates in memory as it jumps from cell to cell are scale-invariant, i.e., each cluster contains, statistically, the same number of subclusters, and so do the subclusters.'

In powder science, the use of scanning electron microscopy has given quantitative information on the pore structure of catalysts. Fractally-based computer-graphic images can now be produced that are similar to the real images (Mann & Wasilewski, 1990). Another example also involves the generation of images: the simulation of 'irregularly shaped particles whose perimeters are fractal.' (Gentry, 1990). This branch of mathematics is useful for producing realistic images, particularly in the life sciences. However, the detail of fractal geometry need not be invoked in dealing with the quantitative aspects of the shape of particle outlines, despite what is often written. Serra (1982), after a succinct, illustrated account of how the Van Koch 'snowflake' is produced, suggests that, if the reader finds the spikiness pathological and too far removed from reality, he should consider some scanning electron micrographs of clay given later in his book. These scans do indicate much roughness, but the outlines do not. In this case, it is possible that transmission electron microscopy would have provided outlines more difficult to characterise.

Surface textures as revealed by video have been classified using resolution analysis. The area of the grey-level surface was measured at several resolutions. This area decreased at coarser resolutions, since fine details that contributed to the area disappeared (Peleg et al., 1984).

A particle surface may well be fractal, and seems often to be so (Pfeifer & Avnir, 1983; Avnir et al., 1983; Avnir et al., 1985; Pines et al., 1988, Schmidt et al., 1989), but the image of the outline that can be obtained in practice rarely is. One published example was cited at the beginning of this chapter. Another is the electronically processed scanning electron micrograph of a coffee agglomerate reproduced in Fig. 6.12. This Figure was used by Normand & Peleg (1986) to illustrate how to obtain a fractal dimension by image processing; the outline is quite rounded.

In a fastidious review of Kaye's book on fractals, Meakin (1991) refers the powder technologist to other texts for precise guidance on the niceties of fractal mathematics; particular mention is made of the book by Feder (1988). At present, if particle outlines or silhouettes are involved, the powder technologist does not need to invoke fractal mathematics. Sensible developments of resolution analysis will provide useful help with silhouette characterisation; but expectations should not be raised too high, even though Kaye (1991), for example, is still making some progress.The care necessary is well described by Graf (1991), who draws particular attention to the problems of resolution. This author writes that fractal information is highly sensitive to resolution limits, and fractal interpretations without consideration of resolution can be misleading. As an example, Graf takes the outline of a lunar-soil grain digitised with 6417 data points (Fig. 6.13(a)). Three 'Richardson diagrams' are presented, with the limits of resolution such that the smallest stride lengths are, respectively, 0.028 x (maximum projected length), 0.003 x (maximum projected

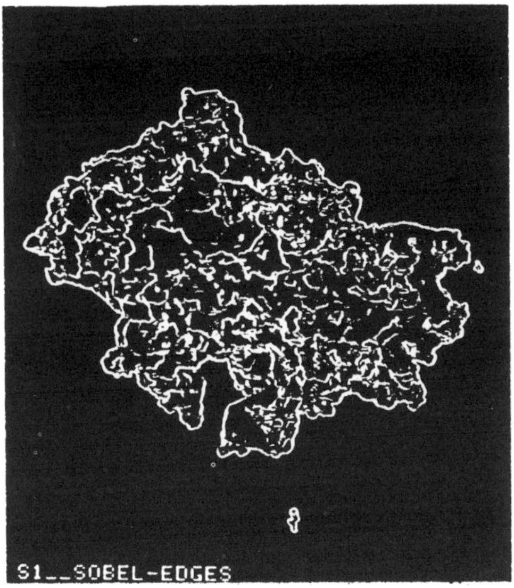

Fig. 6.12 Plot of a digitised image of a coffee granule
(Reproduced from M. D. Normand & M. Peleg, *Powder Technol.*, 1986, **45**, pp.271-275, by permission of the author and Elsevier Sequoia S.A.)

length) and 0.0008 x (maximum projected length). These are reproduced as Figs. 6.13(b), 6.13(c) and 6.13(d), with the author citing one, two and three linear portions. Graf does not appear to question the use of the term 'fractal' when faced with Fig. 6.13(d) and the information this provides about 'self-similarity', nor do Flook & Kaye when using Richardson plots to obtain structural and textural fractal lines. Lovejoy (1982), mentioned above, studied the area-perimeter relation for clouds; he plotted log(area) against log(perimeter) and found linearity over six orders of magnitude. The reports in the powder literature are generally of measurements over two orders of magnitude or less. Some of the work on structure and texture is from measurements extending over less than one order of magnitude.

This Chapter would not be complete without a note about another aspect of the effect of the scale of scrutiny on the apparent length of a wiggly line. Two-dimensional projections and, particularly, sections as viewed through a microscope can be frustrating when information is wanted about the structure of the three-dimensional object from which the section, say, has come. In a chapter entitled 'Stereological Analysis of Particle Characteristics' in a text on testing and characterisation of powders and fine particles (Beddow and Meloy, 1980), Underwood writes:

'Stereology is a fairly recent term, introduced about 15 years ago, and means the study of the three-dimensional structure of a material from two-dimensional sections

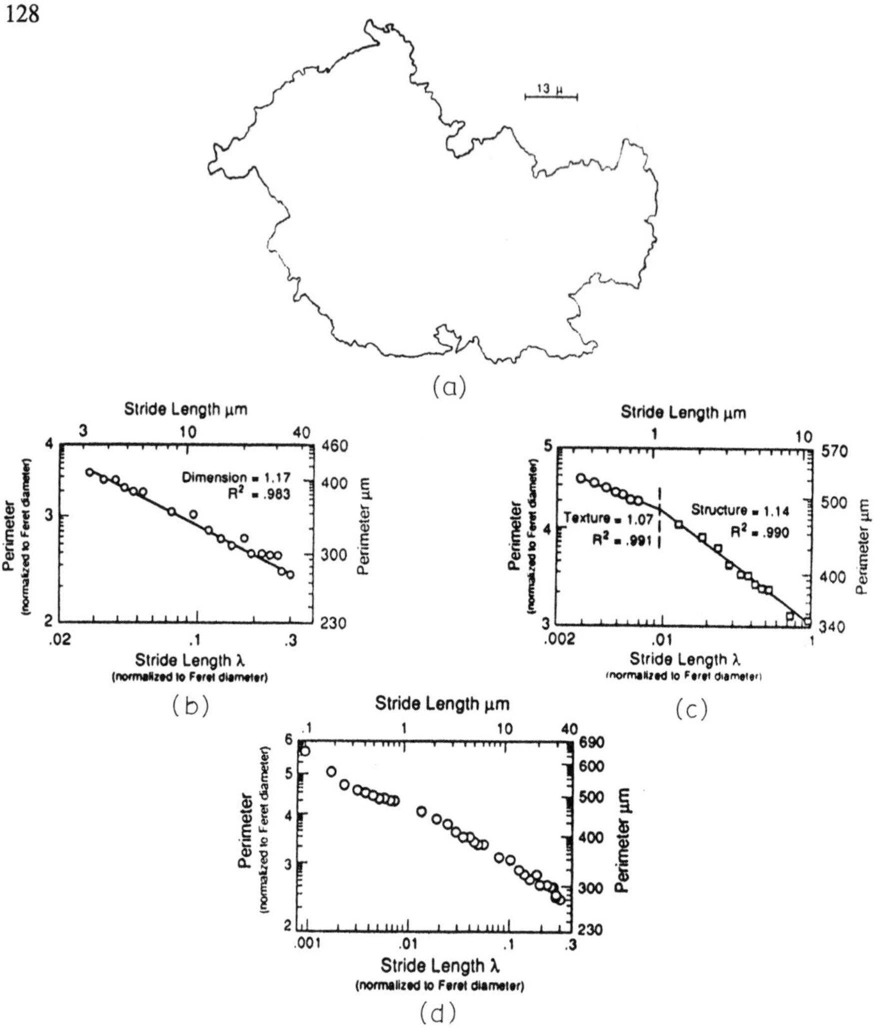

Fig. 6.13 (a) digitised particle outline; (b), (c), (d) Richardson plots using smaller and smaller normalised stride lengths
(Reproduced from J. C. Graf, *Powder Technol.*, 1991, **67**, pp.83-85, by permission of the author and Elsevier Sequoia S.A.)

or projections. Stereology is mainly concerned with the geometrical probabilities, although statistics, topology and projection geometry also contribute to the overall quantitative characterization.'

Underwood has been an important worker in stereology, and his book on the subject is still of considerable influence (1970). Despite his studies and those of many others, it appears to have come as a surprise that the values obtained by stereological techniques are often dependent on the magnification used. An example, quoted by Weibel (1979) and repeated in a similar account by Flook (1982b), is the estimation of the alveolar surface area of human lungs : approximately $80m^2$ by light microscopy and approximately $140m^2$ by electron microscopy. According to Weibel, there are other examples in the stereology literature where serious discrepancies arise when the basic information has been obtained at different magnifications. As Weibel writes, 'this effect has indeed been known to cartographers for a long time.'

This Chapter began by pointing out that resolution analysis quantifies the nature of the line defining the particle outline and not the shape of the outline. This may seem obvious, but Austin (1990), for example, referred to in Chapter 4 as developing a new, single-number classification, states that he did so partly because fractal analysis, as described by Kaye, did not distinguish between outlines clearly very different in shape. This is an example of the need to choose a quantifying system suitable for the task in hand. Another example is provided by an account of the measurement of the shape of polymer powders (Butters, 1985). The shape information required for this task was so rudimentary that even harmonic analysis, restricted to the first few terms, was found to be uneconomic in time and labour. The cost restraint of outline-shape measurement is touched on in the next, concluding, Chapter.

7 Conclusion

The literature concerning powder-particle outlines does not include a 'best way' of quantifying outline shape: the adage 'horses for courses' is applicable. On the academic level, there is a need to tidy up the terminology and the formal definitions of some of the measures used in various disciplines. At the practical level, the greater use of existing statistical techniques would provide more power from combinations of methods of quantification. For example, in palaeontology there were several cases of the use of clustering techniques soon after the arrival of economic computer facilities made this a feasible approach to establishing patterns of similarity (Valentine & Peddicord, 1967; Melle & Buzas, 1968). This was not the situation in powder science and technology.

There has been an accelerating advance in the quality of the facilities available for the employment of established methods of quantification, particularly in throughput; this could improve the reliability of the statistics concerned, provided always that trustworthy sampling is used.

The possible future of image analysis has been assessed in an authoritative manner by Jarvis (1988); this was noted in the Introduction. The progress in supplying fast computer hardware at a reasonable cost should help more researchers to pursue the ideas reviewed in this book. The interest in the mathematics of shape in so many branches of science and technology may provide much of the background support that is desirable. It will be clear, however, that careful thought should be given to research objectives before embarking on the purchase of equipment. A valuable article on this aspect of image analysis was published a few years ago (Moss & Howard, 1988). Although short and now a little dated, the article is still relevant and should be noted.

There are continuing developments in the equipment available for the examination of individual particles. For many years this has been so for microscopes, but now there is increased flexibility and availability of special features. These features, for example computer-controlled stages for the optical microscope and digital video systems for electron microscopes, have changed the general attitude to microscopy.

The recent introduction of confocal laser scanning microscopy could have an impact on the study of powder particles. This new technique bridges the gap between light and electron microscopy. It is true that floor vibration must be low and that there is a significant financial investment involved, but no longer need high magnification, good resolution and depth of field be associated with the examination of a limited number of particles. The figures for performance, to be found in the commercial literature, include focus positioning with better than 0.1μm precision, 200nm resolution, and 16000-time zoom magnification.Possibly more important in the longer term is the fact that focus precision and electronic image storage permit three-dimensional reconstruction. This, for powder particles, may be the development needed to permit substantial advances relating bulk-powder behaviour to particle shape rather than to particle-outline shape.

A need to examine individual particles to obtain their characteristics in their own right can arise, for example, with abrasives or paint pigments in industry or in mathematical modelling in academic studies. This can justify quite substantial investment, but the major commercial drive is almost certainly to optimise powder processing and product performance.

Part of this optimisation can come from introducing a change in powder particle shape or, in some instances, from achieving consistency of shape. A list of some technological parameters and powder properties that may be affected by particle shape is given in Table 7.1. but the time required for the necessary experimentation must be carefully considered when estimating a rate of progress for a research project.

Table **7.1** *–Physicomechanical properties of powders that depend in some degree on particle shape.*

Agglomeration
Aggregation
Bulk density
Cohesivity
Combustion
Compressibility
Explosion
Flowability
Gasification
Packing
Permeability
Porosity
Reactivity
Segregation
Shear behaviour
Suspension stability

So far as powder technology is concerned, there are essentially three research activities that can be identified in this book. They may be grouped in this way:

Group A The application of mathematics to two-dimensional shape.

Group B The application of the findings of group A to particle systems that the researchers are presented with, cannot alter, but wish to understand.

Group C The application of the findings of groups A and B by researchers able to set up their particle systems, which they may wish to manipulate in a predictable and quantifiable manner.

The time scales to produce useful knowledge in the appropriate group increase from group A through to group C and, it would seem, the need for collaborative research also increases. Progress in group C, even with a restricted range of powders and properties, could require substantial effort unless the research be considered very long term. The determination of a reliable measured value of almost any sort for a bulk powder is generally a slow process because of the number of replicates necessary. Moreover, skill, care and experience are required if the standard deviation for the measurement is to be acceptable for correlation purposes. It should not be overlooked that satisfactory bulk-powder experimentation is as difficult to carry out as the more showy quantification of the shape of powder-particle outlines.

An interesting, relatively early, example of what might be achieved with powders was concerned with particle shape and the strength properties of dense asphalt mixes (Lees, 1964b). Partly because of the particle sizes concerned (approximately 5mm-30mm), the development of the research is easily followed and the need for further and wider-ranging experimentation becomes clear. An introductory sentence from the paper is perhaps a proper closure for this book:

'While engineers, whose job is the day-to-day production or use of aggregate, cannot be blamed for taking the view that the detailed description of shape is too lengthy a procedure for routine analysis, no such excuse is permissible for those working in the various fields of research, on whose results the practising engineer ultimately and increasingly depends.'

8 References

Adamson, A. W. (1960), 'Physical Chemistry of Surfaces'. Interscience, N.Y.

Ahlberg, J. H., Nilson, E. N., & Walsh, J. L. (1967), 'The Theory of Splines and Their Application'. Academic Press, N.Y.

Allen, T. (1990), 'Particle Size Measurement', 4th edn. Chapman & Hall, London.

Allen, T. & Khan, A. A. (1970), 'Critical evaluation of powder sampling procedures', *Chem. Eng.*, May, pp.CE 108-112

Allen, T., & Baudet, M. G. (1977), 'The limits of gravitational sedimentation', *Powder Technol.*, **18**, pp.131-138

Andrews, D. F. (1972), 'Plots of high-dimensional data', *Biometrics*, **28**, pp125-136

Anstey, R. L., & Delmet, D. A. (1973), 'Fourier analysis of zooecial shapes in fossil tubular bryozoans', *Bull. Geol. Soc. Am.*, **84**, pp.1753-1764

Aschenbrenner, B. C. (1955), 'A photogrammetric method for the tridimensional measurement of sand grains', *Photogramm. Eng.*, **21**, pp.376-382.

Aschenbrenner, B. C. (1956), 'A new method of expressing particle sphericity', *J. Sedim. Petrol.*, **26**, pp.15-31

Ashby, E. (1948), 'Studies in the morphogenesis of leaves. 1 – An essay on leaf shape', *New Phytol.*, **47**, pp.153-176

Austin, L. G., Yekeler, M., Dumm, T. F., & Hogg, R. (1990), 'The kinetics and shape factors of ultrafine dry grinding in a laboratory tumbling ball mill', *Part. Part. Syst. Charact.*, **7**, pp.242-247

Avnir, D., Farin, D., & Pfeifer, P. (1983), 'Chemistry in noninteger dimensions between two and three. 2 – Fractal surfaces of adsorbents', *J. Chem. Phys.*, **79**, pp.3566-3571

Avnir, D., Farin, D., & Pfeifer, P. (1985), 'Surface geometric irregularity of particulate materials: the fractal approach', *J. Colloid Interface Sci.*, **103**, pp.112-123

Baker, H. W. (1976), 'Environmental sensitivity of submicroscopic surface textures on quartz sand grains – a statistical evaluation', *J. Sedim. Petrol.*, **46**, pp.871-880

Barnsley, M. F. (1988), 'Fractals Everywhere'. Academic Press, Orlando

Barrett, P. J. (1980), 'The shape of rock particles, a critical review', *Sedimentology*, **27**, pp.291-303

Batty, M. (1992), 'Fractals', *Geog. Mag., Lond.*, **64**, (5), pp.33-36; (6), pp.34-36; (7), pp.35-37

Beddow, J. K. (1980), 'Particulate Science and Technology'. Chemical Publishing, N.Y.

Beddow, J. K. (1983), 'Morphological analysis, ten years of progress', *Partic. Sci. & Technol.*, **1**, pp.101-123

Beddow, J. K., & Meloy, T., (1980), 'Testing and Characterization of Powders and Fine Particles'. Heyden, London

Beddow, J. K., & Philip, G. (1975), 'On the use of a Fourier analysis technique for describing the shape of individual particles', *Planseeber. Pulvermet.* **23,** pp.3-14

Beddow, J. K., Nasta, M. D., & Philip, G. C. (1980), 'Characteristics of particle signatures'. *In* 'Testing and Characterization of Powders and Fine Particles' (J.K. Beddow & T.P. Meloy, Eds.). Heyden, London

Beddow, J. K., Philip, G. C., & Vetter, A. F. (1977), 'On relating some particle profile characteristics to the profile Fourier coefficients', *Powder Technol.,* **18,** pp.19-25

Benson, R. H. (1967), 'Muscle-scar patterns of Pleistocene (Kansan) ostracodes'. *In* 'Essays in Palaeontology and Stratigraphy' (C. Teichert & E. L. Yochelson, Eds.). Kansas Univ. Press, Lawrence.

Berry, M. V., & Lewis, Z. V. (1980), 'On the Weierstrass-Mandelbrot fractal function', *Proc. R. Soc.,* **A370,** pp.459-484

Bliss, C. I. (1958), 'Periodic regression in biology and climatology'. Bulletin 615, The Connecticut Agricultural Experiment Station

Boyce, R. R., & Clark, W. A. V. (1964), 'The concept of shape in geography,' *Geog. Rev.,* **54,** pp.561-572

Brady, R. M., & Ball, R. C. (1984), 'Fractal growth of copper electrodeposits', *Nature, Lond., 309,* pp.225 -229

Brill, E. V. (1968), 'Character recognition via Fourier descriptors'. Wescon Tech. Papers Session 25/3, 'Qualitative pattern recognition through image shaping', pp.1-10

B.S. 2955 (1965) 'Glossary of Terms Relating to Powders'. British Standards Institution, London

B.S. 3406. Part 1 (1986), 'British Standard Methods for Determination of Particle Size Distribution. Part 1 – Guide to Powder Sampling'. British Standards Institution, London

Bunday, B. D. (1984), 'Basic Optimisation Methods'. Arnold, London

Bunde, A., & Havlin, S. (Eds.) (1991). 'Fractals and Disordered Systems'. Springer-Verlag, Berlin, Heidelberg, Germany

Burrough, P. A. (1981), 'Fractal dimensions of landscapes and other environmental data', *Nature, Lond.,* **294,** pp.240-242

Butler, P. B., Haworth, M. E., Elban, W. L., & Coyne, P. J. (1990), 'Particle morphology characterization of quasi-statically compacted sucrose powder', *Powder Technol.,* **62,** pp.171-181

Butters, G. (1985), 'Approaches to measuring the particle shape of polymer powders', *In* 'Particle Size Analysis' (P. J. Lloyd, Ed.). Wiley, Chichester

Byerly, G. R., Mrakovich, J. V. & Malcuit, R. J. (1975), 'Use of Fourier shape analysis in zircon petrogenetic studies', *Bull. Geol. Soc. Am.,* **86,** pp.956-958

Cadle, R. D. (1975), 'The Measurement of Airborne Particles'. Wiley, N.Y.

Cailleux, A., (1947), 'L'indice d'émousse: définition et première application', *C. R. Séan. Soc. Géol. Fr.,* **17,** pp.250-252

Champeney, D. C. (1973), 'Fourier Transforms and their Physical Applications'. Academic Press, London

Champeney, D. C. (1985), 'Fourier Transforms in Physics'. Hilger, Bristol

Champeney, D. C. (1987), 'A Handbook of Fourier Theorems'. University Press, Cambridge

Charlier, R., & Goosens, W. (1970/1), 'Sampling a heterogeneous powder using a spinning riffler', *Powder Technol.,* **4,** pp.351-359

Cheng, D. C.-H., & Sutton, H. M. (1971), 'Absolute determination of particle size and shape', *Nature, Lond.*, **232,** pp.192-193

Chermant, J. L., & Coster, M. (1978), 'Fractal object in image analysis'. Proc. Int. Symp. on Quantitative Metallography, Florence, pp.125-137

Chermant, J. L. & Coster, M. (1979), 'Quantitative fractography', *J. Mater. Sci.*, **14,** pp.509-534

Chernoff, H. (1973), 'The use of faces to represent points in k-dimensional space graphically', *J. Am. Statist. Ass.*, **68,** pp.361-368

Church, T. (1968/69), 'Problems associated with the use of the ratio of Martin's diameter to Feret's diameter as a profile shape factor', *Powder Technol.*, **2,** pp.27-31

Clark, M. W. (1981), 'Quantitative shape analysis: a review', *Math. Geol.*, **13,** pp.303-320

Clark, M. W., & Clark, I. (1976), 'A sedimentological pattern recognition problem'. *In* 'Quantitative Techniques for the Analysis of Sediments' (D. F. Merriam, Ed.). Pergamon, Oxford, pp.121-141

Clark, N. N. (1986a) 'Three techniques for implementing digital fractal analysis of particle shape', *Powder Technol.*, **46,** pp.45-52

Clark, N. N. (1986b), 'Fractal harmonics and rugged materials', *Nature, Lond.*, **319,** p.625

Clark, N. N. (1987), 'A new scheme for particle shape characterization based on fractal harmonics and fractal dimensions', *Powder Technol.*, **51,** pp.243-249

Cochran, W. G. (1977), 'Sampling Techniques', 3rd edn., Wiley, N.Y.

Cosgriff, R. L. (1960), 'Identification of shape', Report 820-11 (AD-254,792), The Ohio State University Research Foundation

Coster, M., & Deschanvres, A. (1978), 'Fracture, objet fractal et morphologie mathématique', *Pract. Métallogr.*, **8,** pp.61-73

Cox, E. P. (1927), 'A method of assigning numerical and percentage values to the degree of roundness', *J. Paleont.*,**1,** pp.179-183

D'Arcy Thompson, A. W. (1942), 'On Growth and Form'. University Press, Cambridge

D'Arcy Thompson, A. W. (1961), 'On Growth and Form', Abridged edn. (J. T. Bonner, Ed.). University Press, Cambridge

Davies, C. N. (1962), 'Measurement of particles', *Nature, Lond.*, **195,** pp.768-770

Davies, C. N. (1964), 'Shape of small particles', *Nature, Lond.*, **201,** pp.905-907

Davies, K. W. (1979), Personal communication

Davies, K. W., & Hawkins, A. E. (1979), 'Harmonic descriptors of particle shape: their role in discrimination', Proc. 2nd Eur. Symp. on Particle Characterisation, Nuremberg, pp.582-590

Davies, R. (1968/69), 'Settling of particles in suspension at high concentrations', *Powder Technol.*, **2,** pp.43-51

Davies, R. (1975), 'A simple feature-space representation of particle shape', *Powder Technol.*, **12,** pp.111-124

Davies, R., & Karuhn, R. (1973), 'The use of triaxial measurements to describe the sedimentation, transport, and abrasion behaviour of particles', Proc. Harold Heywood Memorial Symp., University of Loughborough

Davis, J. C. (1973), 'Statistics and Data Analysis in Geology'. Wiley, N.Y.

De Silva R. L., & Creasy, D. E. (1983), 'Shape factors for barium chromate crystals in the sieve size range', *Powder Technol.*, **35,** pp.181-184

Drescher, S., Heidenreich, E., & Müller (1990), 'Technologically relevant particle shape analysis', *Part. Part. Syst. Charact.*, **7,** pp.30-35

Duda, R. O., & Hart, P. E. (1973), 'Pattern Classification and Scene Analysis'. Wiley-Interscience, N.Y.

Editorial (1991), 'Physical graffiti', *Phys. World*, **4,** p.18

Ehrlich, R., & Weinberg, B. (1970), 'An exact method for characterisation of grain shape', *J. Sedim. Petrol.*, **40,** pp.205-212

Ehrlich, R., Brown, P. J., Yarus, J. M., & Przygocki, R. S. (1980a), 'The origin of shape frequency distributions and the relationship between size and shape', *J. Sedim. Petrol.*, **50,** pp.475-483

Ehrlich, R., Brown, P. J., Yarus, J. M., & Eppler, D. T. (1980b), 'Analysis of particle morphology data', *In* 'Advanced Particulate Morphology' (J. K.Beddow & T .P. Meloy, Eds.). CRC Press, Boca Raton, p.101-119

Erens, R. (1991), 'Research Libraries in Transition; Academic Perceptions of Recent Developments in University and Polytechnic Libraries'. British Library Research & Development Department, Boston Spa

Fairs, G. L. (1943), 'The use of the microscope in particle size analysis', *Chem. Ind.*, **62,** pp.374-378

Falconer, K. (1990), 'Fractal Geometry: Mathematical Foundations and Applications'. Wiley, Chichester

Farin, D., Peleg, S., Yavin, D., & Avnir, D. (1985a), 'Applications and limitations of boundary-line fractal analysis of irregular surfaces: proteins, aggregates and porous materials', *Langmuir,* **1,** pp.399-407

Farin, D., Volpert, A., & Avnir, D. (1985b), 'Determination of adsorption conformation from surface resolution analysis', *J. Am. Chem. Soc.*, **107,** pp.3368-3370

Feder, J. (1988), 'Fractals'. Plenum Press, New York

Finch, H., & North, C. (1991), 'The Research Process: The Library's Contribution'. British Library Research & Development Department, Boston Spa

Fischer, G. (1933), 'Die Petrographie der Grauwacken', *Jahrb. d. Preuss. Geol. Landesanst.*, **54,** pp.322-323

Flemming, N. C. (1965), 'Form and function of sedimentary particles', *J. Sedim. Petrol.*, **35,** pp.381-390

Flook, A. G. (1978), 'The use of dilation logic on the Quantimet to achieve fractal dimension characterisation of textured and structured profiles', *Powder Technol.*, **21,** pp.295-298

Flook, A. G. (1979), 'The characterization of textured and structured particle profiles by the automated measurement of their fractal dimensions'. Proc. 2nd Eur. Symp. on Particle Characterisation, Nuremberg, pp.591-599

Flook, A. G. (1982a), 'Fourier analysis of particle shape'. *In* 'Particle Size Analysis 1981' (N.G. Stanley-Wood and T. Allen, Eds.). Wiley-Heyden, London

Flook, A. G. (1982b), 'Fractal dimensions: their evaluation and their significance in stereological measurements', *Acta Stereol.*, **1,** pp.79-87

Fong, S.-T., Beddow, J. K., & Vetter, A. F. (1979a), 'A refined method of particle shape representation', *Powder Technol.*, **22,** pp.17-21

Fong, S.-T., Beddow, J. K., & Vetter, A. F. (1979b), 'Acid-zinc reaction predictably affects particle morphology', *Powder Technol.*, **23,** pp.187-190

Fox, R. D., & Brazee, R. D. (1971), 'Numerical methods of describing leaf shapes', *J. Agric. Eng. Res.*, **16,** pp.353-363

Free, E. E. (1911), 'The movement of soil material by wind', *Bull. Div. Soils, U.S. Dep. Agric.*, **68**

Freeman, H. (1961), 'On the encoding of arbitrary geometric configurations', *IRE Trans. Electron. Compon.*, **EC-10,** pp.260-268

Freshwater, D. C. (1987), 'Harold Heywood – The Mann *(sic)* and his work', *Powder Technol.*, **50,** pp.295-299

Full, W. E., & Ehrlich, R. (1982), 'Some approaches for location of centroids of quartz grain outlines to increase homology between Fourier amplitude spectra', *Math. Geol.*, **14,** pp.43-55

Gentry, J. W. (1990), 'The use of simulated fractals in particle characterization', *Part. Part. Syst. Charact.*, **7,** pp.87-92

Gevirtz, J. L. (1976), 'Fourier analysis of bivalve outlines: implications on evolution and autecology', *Math. Geol.*, **8,** pp.151-163

Gibbard, D. W., Smith, D. J., & Wells, A. (1972), 'Area sizing and pattern recognition on the Quantimet 720', *Microscope,* **20,** pp.37-50

Gotoh, K. (1979), 'Shape characterization of two-dimensional forms', *Powder Technol.*, **23,** pp.131-134

Goudie, A. (1981), 'Geomorphological Techniques', George Allen & Unwin, London

Graf, J. C. (1991), 'The importance of resolution limits to the interpretation of fractal descriptions of fine particles', *Powder Technol.*, **67,** pp.83-85

Gy, P. M. (1967), 'L'Échantillonnage des Minerais en Vrac – Vol. 1', Société de l'Industrie Minérale, Saint-Etienne

Gy, P. M. (1971), 'L'Échantillonnage des Minerais en Vrac – Vol. 2', Société de l' Industrie Minérale, Saint-Etienne

Gy, P. M. (1976), 'The sampling of particulate materials – a general theory', *Int. J. Miner. Process.*, **3,** pp.289-312

Gy, P. M. (1982), 'Sampling of Particulate Materials: Theory and Practice', 2nd edn., Elsevier, Amsterdam

Gy, P. M. (1983), 'Sampling errors – they may deprive the analytical results of any meaning', *Analusis,* **11,** pp.413-440

Gy, P. M. (1986), 'The analytical and economic importance of correctness in sampling', *Analyt. Chim.*, **190,** pp.13-23

Gy, P. M, & Marin, L., (1978), 'Unbiased sampling from a falling stream of particulate material', *Int. J. Miner. Process.* **5,** pp.297-315

Hagerman, T. H. (1933), 'Some lithological methods for determination of stratigraphic horizons', *Proc. World Petroleum Congress, London,* **1,** pp.257 -259

Hammond, D. (1941), 'The expression of genes for leaf shape in *Gossypium hirsutum L.* and *Gossypium arboreum L.*', *Am. J. Bot.*, **28,** pp.124-134

Harbaugh, J. W., & Preston, F. W. (1966), 'Fourier series analysis in geology'. *In* 'Computers and Computer Applications in Mining and Exploration'. School of Mines, Univ. of Arizona, pp.R1-R46

Hartley, H. O. (1949), 'Tests of significance in harmonic analysis', *Biometrika,* **36,** pp.194-201

Hatton, T. A. (1978), 'Representative sampling of particles with a spinning riffler. A stochastic model', *Powder Technol.*, **19,** pp.227-233

Hausner, H. H. (1966), 'Characterization of the powder particle shape', *Planseeber. Pulvermet.*, **14,** pp.75-84

Hausner, H. H. (1967), 'Characterisation of the powder particle shape'. *In* 'Particle Size Analysis'. Society for Analytical Chemistry, London, pp.20-28

Hawkins, A. E. (1980), 'A reconsideration of rotating vector particle shape description'. Proc. Tech. Sessions, Powder Europa, Wiesbaden

Hawkins, A. E. (1987), 'The probable number of particles in contact when spread on a microscope slide', *Part. Charact.*, **4,** pp.20-21

Hawkins, A. E. (1988), 'Demonstrating the sampling of heterogeneous, free-flowing, segregating powders', *Part. Part. Syst. Charact.*, **5,** pp.23-28

Hawkins, A. E., & Davies, K. W. (1984), 'Sampling powders for shape description', *Part. Charact.*, **1,** pp.22-27

Hayes, B., & Fitzke, F. (1987), 'Design and use of a semi-automatic image analysis system', *Lab. Pract.*, **36,** pp.78-79

Herdan, G. (1960), 'Small Particle Statistics', 2nd edn., Butterworths, London

Heyd, A., & Dhabhar, D. (1979), 'Particle shape effect on caking of coarse granulated antacid suspensions', *Drug Cosmet. Ind.*, **125,** pp.42 -45, 146-147

Heywood, H. (1937), 'Numerical definitions of particle size and shape', *Chem. Ind.*, **56,** pp.149-154

Heywood, H. (1938), 'Measurement of the fineness of powdered materials', *Proc. Inst. Mech. Eng.*, **140,** pp.257-347

Heywood, H. (1945), 'A comparison of methods of measuring microscopical particles', *Trans. Inst. Min. Met.*, **4,** pp.391-404

Herrmann, H. J., Mantica, G., & Bessis, D. (1990), Space-filling bearings', *Phys. Rev. Lett.*, **65,** pp.3223-3226

Holt, C. B. (1981), 'Characterisation of the shape of particles produced by crushing using harmonic series analysis', Proc. Powtech., Birmingham. I.Chem.E. Symposium Series No. 63, pp.D2/H/1-13

Horn, L. H., & Bryson, R. A. (1960), 'Harmonic analysis of the annual march of precipitation over the United States', *Ann. Ass. Am. Geog.*, **50,** pp.157-171

Horton, R. E. (1932), 'Drainage-basin characteristics', *Trans. Am. Geophys. Union,* **14,** pp.350-361

Huller, D., & Weichert, R. (1982), 'Quantitative shape analysis of particle contours'. *In* 'Particle Size Analysis 1981' (N.G. Stanley-Wood & T. Allen, eds.). Wiley-Heyden

Humphries, D. W. (1961), 'A non-laminated miniature sample splitter', *J. Sedim. Petrol.*, **31,** pp.471-473

Institution of Mechanical Engineers (1991). Seminar. 'The Sampling of Bulk Materials'. May, London

Ivry, B. (1990), 'The beauty of messiness', *Newsweek,* 26th March, p.56

Jain, A. K., & Dubes, R. C. (1988), 'Algorithms for Clustering Data'. Prentice-Hall, Englewood Cliffs

Jarvis, L. R. (1988), 'Microcomputer video image analysis', *J. Microsc.*, **150,** pp.83-97

Jarvis, R. S. (1976), 'Classification of nested tributary basins in analysis of drainage basin shape', *Water Resources Res.*, **12,** pp.1151-1164

Jones, S. (1983), 'The problem of closure in the Zahn-Roskies method of shape description', *Powder Technol.*, **34,** pp.93-94

Kaesler, R. L., & Waters, J. A. (1972), 'Fourier analysis of the ostracode margin'. *Bull. Geol. Soc. Am.*, **83,** pp.1169-1178

Kaye, B. H. (1962), 'An investigation into the relative efficiency of different sampling procedures', *Powder Met.*, **9,** pp.213-234

Kaye, B. H. (1964), 'Efficient sample reduction by means of a riffler sampler', *Monogr. Soc. Chem. Ind.*, **18,** pp.159-167

Kaye, B. H. (1968/69), 'Some aspects of the efficiency of statistical methods of particle size analysis', *Powder Technol.*, **2**, pp.97-110

Kaye, B. H. (1978a), 'Characterization of the surface area of a fine particle profile by its fractal dimension'. *In* 'Particle Size Analysis' (M. J. Groves, Ed.). Heyden, London

Kaye, B. H. (1978b), 'Specification of the ruggedness and/or texture of a fine particle profile by its fractal dimension', *Powder. Technol.*, **21**, pp.1-16

Kaye, B. H. (1981), 'Direct Characterization of Fine Particles'. Wiley, New York

Kaye, B. H. (1984), 'Multifractal description of a rugged fineparticle profile', *Part. Charact.*, **1**, pp.14-21

Kaye, B. H. (1986), 'The description of two-dimensional rugged boundaries in fineparticle science by means of fractal dimensions', *Powder Technol.*, **46**, pp.245-254

Kaye, B. H. (1989), 'A Random Walk Through Fractal Dimensions'. VCH, Weinheim

Kaye, B. H. (1991), 'Characterizing the structure of fumed pigments using the concepts of fractal geometry', *Part. Part. Syst. Charact.*, **8**, pp.63-71

Kaye, B. H., & Clarke, G. G. (1986), 'Fractal description of extra terrestrial fineparticles', *Part. Charact.*, **4**, pp.143-148

Kaye, B. H., Clark, G. G., Leblanc, J. E., & Trottier, R. A. (1986), 'Image analysis procedures for characterizing the fractal dimension of fineparticles', Proc. 1st World Congr. on Particle Technology, Nuremberg, pp.17-29

Kendal, Sir M. (1975), 'Multivariate Analysis'. Charles Griffin, High Wycombe

Kennedy, S. K. (1985), 'Microcomputer image acquisition and shape and size analysis with applications in sedimentology', *Compass of Sigma Gamma Epsilon*, **62**, pp.205- 210

Kennedy, S. K., & Lin, W.-H. (1986), 'Fract – a Fortran subroutine to calculate the variables necessary to determine the fractal dimension of closed forms', *Comput. Geosci.*, **12**, pp.705-712

Kjems, J. K. (1991), 'Fractals and experiments'. *In* 'Fractals and Disordered Systems' (A. Bunde & S. Havlin, Eds.). Springer-Verlag, Berlin, Heidelberg

Krumbein, W. C. (1941), 'Measurement and geological significance of shape and roundness of sedimentary particles', *J. Sedim. Petrol.*, **11**, pp.64-72

Krumbein, W. C., & Graybill, F. A. (1965), 'An Introduction to Statistical Models in Geology', McGraw-Hill, N.Y.

Krumbein, W. C., & Pettijohn, F. J. (1938), 'Manual of Sedimentary Petrography', D. Appleton-Century, N.Y./London

Lam, K. K., & Newton, J. M. (1991), 'Investigation of applied compression on the adhesion of powders to a substrate surface', *Powder Technol.*, **65**, pp.167-175

Lancaster, H. O. (1970), 'Statistical control in haematology'. *In* 'Random Counts in Scientific Work – Vol.2' (G. P. Patil, Ed.). Pennsylvania State University Press, University Park

Lee, Y., Beddow, J. K., Vetter, A. F., and Lenth, R. (1980), 'Morphological analysis of fine particle mixtures', *Powder Technol.*, **25**, pp.137-145

Lees, G. (1964a), 'A new method for determining the angularity of particles', *Sedimentology*, **3**, pp.2-21

Lees, G. (1964b), 'The measurement of particle shape and its influence in engineering materials', *J. Br. Granite Whinstone Fed.*, **4**, pp.17-38

Lenn, C. P., & Holt, C. B. (1982), 'The use of the Jeffrey-Galion shape sorting table in the production of shape sorted populations of particles'. *In* 'Particle Size Analysis 1981' (N. G. Stanley-Wood & T. Allen, Eds.). Wiley, Chichester

Lennard, P. R. (1990), 'Image analysis for all', *Nature, Lond.*, **347,** pp.103-104

Lieberman, A. (1984), 'Fine particle characterization methods in liquid suspensions'. In 'Particle Characterization in Technology – Vol. 1' (J. K. Beddow, Ed.). CRC Press, Boca Raton

Lindley, D. V. (1947), 'Regression lines and the linear functional relationship', *J. R. Statist. Soc. Suppl.*, **9,** pp.218-244

Lloyd, P. J. (1988), 'The characterisation of particle shape'. *In* 'Particle Size Analysis' (P. J. Lloyd, Ed.). Wiley, Chichester

Lovejoy, S. (1982), 'Area-perimeter relation for rain and cloud areas', *Science,* N.Y., **216,** pp.185-187

Lu, K. H. (1965), 'Harmonic analysis of the human face', *Biometrics,* **21,** pp.491-505

Luerkens, D.W. (1991), 'The Theory and Application of Morphological Analysis: Fine Particles and Surfaces'. CRC Press, Boca Raton

Luerkens, D. W., Beddow, J. K., & Vetter, A. F. (1982), 'Morphological Fourier descriptors', *Powder Technol.,* **31,** pp.209-215

Luerkens, D. W., Beddow, J. K., & Vetter, A. F. (1987), 'Structure and morphology – the science of form applied to particle characterization', *Particle Technol.,* **50,** pp.93-101

Mackey, R. D. (1965), 'The measurement of particle shape', *Civ. Eng. Publ. Wks. Rev.,* **60,** pp.211-214

Maeder, A. E., & Clark, N. N. (1991), 'Harmonic endurance: a new shape descriptor derived from polygonal harmonics', *Powder Technol.,* **68,** pp.137-143

Mandelbrot, B. B. (1967), 'How long is the coast of Britain? Statistical self-similarity and fractional dimension', *Science,* N.Y., **156,** pp.636-638

Mandelbrot, B. B. (1975), 'Les Objets Fractals, Forme, Hasard, Dimension'. Flammarion, Paris

Mandelbrot, B. B. (1977), 'Fractals, Form, Chance and Dimension'. W. H. Freeman, San Francisco

Mandelbrot, B. B. (1982), 'The Fractal Geometry of Nature'. W. H. Freeman, New York

Mann, R., & Wasilewski, M. C. (1990), 'Towards a fractal computer graphic basis for characterisation of catalyst pore structure by image reconstruction', *Trans. Inst. Chem. Eng.,* **68,** pp.177-184

Mark, D. M., & Church, M. (1977), 'On the misuse of regression in earth science', *Math. Geol.,* **9,** pp.63-75

Markwick, A. H. D. (1937), 'The shape of crushed stone and gravel and its measurement, *Chem. Ind.,* **56,** pp.206-213

McCarthy, C. J. (1976), 'Recent advances in the characterization of particulates using automatic evaluation of shape descriptors', *Microstruct. Sci.,* **4,** pp.339-346

McCarthy, D. A., Bernhagen, J., Lin, Y.-C., & Perry, J. D. (1990), 'A rapid preparation technique for leucocytes', *J. Microsc.,* **158,** pp.63-72

Meakin, P. (1991). Book Review: 'A Random Walk Through Fractal Dimensions' by B. H. Kaye, 1989. *Powder Technol.,* **64,** p.287

Medalia, A. I. (1970/71), 'Dynamic shape factors of particles', *Powder Technol.,* **4,** 117- 138

Melle, J. F., & Buzas, M. A. (1968), 'An application of cluster analysis as a method of determining biofacies', *J. Paleont.,* **42,** pp.747-758

Meloy, T. P. (1977a), 'A hypothesis for morphological characterization of particle shape and physiochemical properties', *Powder Technol.,* **16,** pp.233-253

Meloy, T. P. (1977b), 'Fast Fourier transforms applied to shape analysis of particle silhouettes to obtain morphological data', *Powder Technol.,* **17,** pp.27-35

Meloy, T. P. (1980), 'Particle shape characterization: recent developments (Review).' *In* 'Testing and Characterization of Powders and Fine Particles' (J. K. Beddow and T. P. Meloy, Eds.). Heyden, London

Meloy, T. P. (1983), 'Particulate characterization: future approaches'. *In* 'Handbook of Powder Science and Technology' (M.E. Fayed & L. Otten, Eds.). Van Nostrand Reinhold, N.Y.

Meloy, T. P., & Clark, N. N. (1986), 'Shape characterisation of angular particles related to their behavior', Proc. 1st World Congress Particle Technology, Nuremberg, pp.93-103

Meloy, T. P., and Makino, K. (1983), Characterizing residence times of powder samples on sieves', *Powder Technol.*, **36**, pp.253-258

Meltzer, B., Searle, N. H., & Brown, R. (1967), 'Numerical specification of biological form', *Nature, Lond.*, **216**, pp.32-36

Miller, V. C. (1953), 'A quantitative geomorphic study of drainage basin characteristics in the Clinch Mountain Area, Virginia and Tennessee', Columbia Univ. Dept. of Geol., Tech. Rep. No. 3, N.Y.

Modes, S., Lianos, P., & Xenakis, A. (1990), 'Relation of fractal behaviour of luminescence quenching with electric percolation in water-in-oil emulsions', *J. Phys. Chem.*, **94**, pp.3363-3365

Montgomery, J. R. (1968), 'Revolving riffle for subdividing ground magnesium chips', *Anal. Chem.*, **40**, pp.1399-1400

Moss, M., & Howard, V. (1988), 'Look before you leap', *Lab. Pract.*, **37**, pp.15-23

Müller, G. (1967), 'Methods in Sedimentary Petrology'. Part 1 of 'Sedimentary Petrology' by W. v. Engelhardt, H. Füchtbauer & G. Müller (Translated by Hans-Ulrich Schmincke). Hafner, N.Y./London

Murphy, C. H. (1984), 'Handbook of Particle Sampling and Analysis Methods'. Verlag Chemie International, Deerfield Beach

Nakajima, Y., Whiten, W. J. & White, M. E. (1978), 'Method for measurement of particle-shape distribution by sieves'. *Trans. Inst. Min. Met.*, **87**, pp.C194-C203

Needham, L. W., & Hill, N. W. (1935), 'The shape and specific surface of coal particles', *Fuel, Lond.*, **14**, pp.222-230

Newman, T. M. (1989), 'The use of new image storage techniques in microscopy', *Microscopy & Analysis*, July, pp.23-26

Normand, M. D., & Peleg, M. (1986), 'Determination of the fractal dimension of a particle silhouette using image-processing techniques', *Powder Technol.*, **45**, pp.271-275

Orford, J. D., & Whalley, W. B. (1983), 'The use of the fractal dimension to quantify the morphology of irregular-shaped particles', *Sedimentology*, **30**, pp.655-668

Otto, G. H. (1937), 'The use of statistical methods in effecting improvements on a Jones sample splitter'. *J. Sedim. Petrol.*, **7**, pp.101-133

Pahl, M. H., Schädel, G., & Rumpf, H. (1973), 'Zusammenstellung von Teilchenformbescreibungsmethoden', *Aufbereitungs Tech.* **14**, pp.257-264, 672-683, 759-764

Pang, H. M., & Ridgway, K. (1980), 'The extension of the vibratory shape-sorting technique to smaller particle sizes', *J. Pharm., Pharmac. Suppl.*, **32**, pp.19-23

Peixoto, J. P., Saltzman, B., & Teweles, S. (1964), 'Harmonic analysis of the topography along parallels of the earth'. *J. Geophys. Res.*, **69**, pp.1501-1505

Peleg, S., Naor, J., Hartley, R., & Avnir, D. (1984), Multiple resolution texture analysis and classification', *IEEE Trans. Pattern Anal. Mach. Intell.*, **6**, pp.518-523

Perry, R. W., Harris, J. E. C., & Scullion, H. J. (1978), "Difficult' particulates – the application of microscopy and other techniques'. *In* 'Particle Size Analysis' (M. J. Groves, Ed.) Heyden, London

Pettijohn, F. J. (1957), 'Sedimentary Rocks', 2nd edn. Harper & Brothers, N.Y.

Pfeifer, P., & Avnir, D. (1983), 'Chemistry in noniteger dimensions between two and three. 1 – Fractal theory of heterogenous surfaces', *J. Chem. Phys.*, **79**, pp.3558-3565

Pickett, J., Clark, N., & Shuker, S. (1991), 'A re-examination of Piper's shape analysis technique', *Part. Part. Syst. Charact.*, **8**, pp.110-115

Pines, D., Huppert, D., & Avnir, D. (1988), 'The fractal nature of the surfaces of porous silicas as revealed in electronic energy transfer between adsorbates: comparison of three donor/acceptor pairs', *J. Chem. Phys.*, **89**, pp.1177-1180

Piper, D. J. W. (1971), 'The use of the D-Mac Pencil Follower in routine determinations of sedimentary parameters', *In* 'Data Processing in Biology and Geology' (J. L. Cutbill, Ed.). Academic Press, London

Plotnick, R. E. (1986), 'A fractal model for the distribution of stratigraphic hiatuses', *J. Geol.*, **94**, pp.885-890

Polke, R., Schäfer, M., & Scholz, N. (1991), 'Preparation technology for fine particle measurement', *Part. Part. Syst. Charact.*, **8**, pp.1-7

Powers, M. C. (1953), 'A new roundness scale for sedimentary particles', *J. Sedim. Petrol.*, **23**, pp.117-119

Preston, F. W., & Henderson, J. H. (1964), 'Fourier series characterization of cyclic sediments for stratigraphic correlation'. *In* 'Symposium on Cyclic Sedimentation'. *Kansas Geol. Survey Bull.*, **2**, pp.415 -425

Redman, E., Heckman, F. A., & Connolly, J. E. (1978), 'A practical approach to reliable quantitative image analysis of carbon black electron micrographs'. *In* 'Particle Size Analysis' (M. J. Groves, Ed.). Heyden, London

Reilly, S., & Clark, N. N. (1991), 'Computer based general shape description with polygonal harmonics and a roughness index'. *Proc. Inst. Mech. Eng.*, **205**, pp.103-111

Réti, T., & Czinege, I. (1989), 'Shape characterization of particles via generalized Fourier analysis', *J. Microsc.*, **156**, pp.15-32

Richardson, L. F. (1960), 'Statistics of Deadly Quarrels' (Q. Wright & C. C. Lienau, Eds.). Boxwood Press, Pittsburgh

Richardson, L. F. (1961), 'The problem of contiguity: an appendix of statistics of deadly quarrels', *Gen. Syst. Yearb.*, **6**, pp.139-187

Ridgway, K., & Rupp, R. (1969), 'The effect of particle shape on powder properties', J. Pharm., Pharmac. Suppl., **21**, pp.30-39

Ridgway, K., & Rupp, R. (1970/71), 'The mixing of powder layers on a chute: the effect of particle size and shape', *Powder Technol.*, **4**, pp.195-202

Riebel, U., Kofler, V., & Löffler, F. (1991), 'Shape characterization of crystals and crystal agglomerates', *Part. Part. Syst. Charact.*, **8**, pp.48-54

Riley, G. S. (1977), 'In search of an easily measurable particle shape index', *J. Powder & Bulk Solids Technol.*, **1**, pp.34-39

Riley, N A. (1941), 'Projection sphericity', *J. Sedim. Petrol.*, **11**, pp.94-97

Rittenhouse, G. R. (1943), 'A visual method of estimating two-dimensional sphericity', *J. Sedim. Petrol.*, **13**, pp.79-81

Robertson, R. H. S., & Emödi, B. S. (1943), 'Rugosity of granular solids', *Nature, Lond.*, **152**, pp.539-540

Rose, H. E. (1953), 'The Measurement of Particle Size in Very Fine Powders'. Constable, London

Rosenfeld, M. A., & Griffiths, J. C. (1953), 'An experimental test of visual comparison technique in estimating two dimensional sphericity and roundness of quartz grains', *Am. J. Sci.*, **251,** pp.553-585

Rösler, R., Schneider, H. A., & Schuberth, R. (1987), 'Relation between particle shape and profile Fourier coefficients', *Powder Technol.*, **49,** pp.255-260

Rumpf, H. (1990), 'Particle Technology' (Translated by F. A. Bull). Chapman & Hall, London

Russell, R. D., & Taylor, R. E. (1937), 'Roundness and shape of Mississippi River sands', *J. Geol.*, **45,** pp.225-267

Rys, F. S., & Waldvogel, A. (1986), 'Analysis of the fractal shape of severe convective clouds'. *In* 'Fractals in Physics' (L. Pietronero & E. Tosatti, Eds.). Elsevier, Amsterdam

Sankar, P. V., & Krishnamurthy, E. V. (1978), 'On the compactness of subsets of digital pictures', *Comput. Graphics & Image Process.*, **8,** pp.136-143

Schmidt, P. W., Avnir, D., & Lin, J. S. (1989), 'Small angle X-ray scattering study of the fractal morphology of porous silicas', *J. Chem. Phys.*, **90,** pp.5016-5023

Schneiderhöhn, P. (1954), 'Eine vergleichende Studie über Methoden zur quantitativen Bestimmung von Abrundung und Form an Sandkörnern', *Heidelb. Beitr. Miner. Petrogr.*, **4,** pp.172-191

Schwarz, H., & Exner, H. E. (1980), 'The implementation of the concept of fractal dimension on a semi-automatic image analyser', *Powder Technol.*, **27,** pp.207-213

Schwarcz, H. P., & Shane, K. C. (1969), 'Measurement of particle shape by Fourier analysis', *Sedimentology,* **13,** pp.213-231

Scott, K. J. (1984), 'Sedimentation'. *In* 'Handbook of Powder Science and Technology' (M. E. Fayed & L.Otten, Eds.). Van Nostrand Reinhold, N.Y.

Sergeev, Y. M., Spivak, G. V., Sasov, A. Y., Osipov, V. I., & Sokolov, V. N. (1984), 'Quantitative morphological analysis in a SEM-microcomputer system. I – Quantitative shape analysis of single objects', *J. Microsc.*, **135,** pp.1-12

Serra, J. (1982), 'Image Analysis and Mathematical Morphology'. Academic Press, London

Siegrist, H. G., & Wylie, A. G. (1980), 'Characterizing and discriminating the shape of asbestos particles', *Environ. Res.*, **23,** pp.348-361

Singh, P., & Ramakrishnau, P. (1988), 'Morphological classification of copper powders of different origins by Fourier analysis', *Powder Met.*, **31,** pp.287-291

Sorby, H. C. (1880), 'On the structure and origin of non-calcareous stratified rocks', *Q. J. Geol. Soc. Lond.*, **36,** pp.46-92

Sroczynski, S. F. (1990), 'Methods for obtaining high quality stereoscopic images of microscopic objects', *J. Microsc.*, **157,** pp.163-170

Stabler, C. L. (1968), 'Simplified Fourier analysis of fold shapes', *Tectonophys.*, **6,** pp.343-350

Starkey, J., & Simigian, S. (1987), 'Image: a Fortran V program for image analysis of particles', *Comput. & Geosci.*, **13,** pp.37-59

Steinhaus, H. (1954), 'Length, shape and area', *Colloq. Math.*, **3,** pp.1-13

Stentiford, M. J. (1979), 'What about the sample?', *Trans. J. Br. Ceram. Soc.*, **78,** pp.10-12

Stone, R. O., & Dugundji, J. (1965), 'A study of microrelief – its mapping, classification, and quantification by means of a Fourier analysis', *Eng. Geol.*, **1,** pp.55-187

Strahler, A. N. (1958), 'Dimensional analysis applied to fluvially eroded landforms', *Bull. Geol. Soc. Am.*, **69,** pp.279-300

Student (1907), 'On the error of counting with a haemacytometer', *Biometrika,* **5,** pp.351- 360

Sutton, H. M., & Bundalli, N. (1973), 'A suggestion for the exact determination of particle size and shape', *Proc. Soc. Anal. Chem.,* **10,** pp.13-17

Svarovsky, L. (1987), 'Powder Testing Guide'. Elsevier Applied Science, London

Szádeczy-Kardoss, E. von (1933), 'Die Bestimmung der Abrollungsgrades', *Zentbl. Miner. Geol. Paläeont.,* **34B,** pp.389-401

Thaulow, N., & White, E. W. (1971/72), 'General method of dispersing and disaggregating particulate samples for quantitative SEM and optical microscope studies', *Powder Technol.,* **5,** pp.373-384

Thiébaut, D. (1989), 'On the fractal dimension of computer programs and its application to the prediction of the cache miss ratio', *IEEE Trans. Comput.,* **C-38,** pp.1012-1026

Tickell, F. G. (1931), 'The Examination of Fragmental Rocks'. Stanford University Press, California

Tilmann, S. E. (1973), 'The effect of grain orientation on Fourier shape analysis', *J. Sedim. Petrol.,* **43,** pp.867-869

Tsubaki, J., & Jimbo, G. (1979), 'A proposed new characterization of particle shape and its application', *Powder Technol.,* **22,** pp.161-169

Underwood, E. E. (1970), 'Quantitative Stereology'. Addison-Wesley, Reading, Mass.

Underwood, E. E. (1980), 'Stereological analysis of particle characteristics'. *In* 'Testing and Characterization of Powders and Fine Particles' (J. K. Beddow & T. P. Meloy, Eds.). Heyden, London

Valentine, J. W., & Peddicord, R. G. (1967), 'Evaluation of fossil assemblages by cluster analysis', *J. Paleont.,* **41,** pp.502-507

Van Nieuwenhuise, D. S., Yarus, J. M., Przygocki, R. S., & Ehrlich, R. (1978), 'Sources of shoaling in Charleston Harbor; Fourier grain shape analysis', *J. Sedim. Petrol.,* **48,** pp.373-383

Viswanathan, K., Aravamudhan, S., & Mani, B. P. (1984), 'Separation based on shape. Part 1 – recovery efficiency of spherical particles', *Powder Technol.,* **39,** pp.83-91

von Endter, F., & Gebauer, H. (1965), 'Ein einfaches Gerat zur Statistischen Auswertung von Mikroskopischen bzw. Elektronenmikroskopischen aufnahmen', *Optik,* **13,** pp.97- 103

von Rittinger, P. R. (1867), 'Lehrbuch der Aufbereitungskunde'. W. Ernst u. Sohn, Berlin

Wadell, H. (1932), 'Volume, shape and roundness of rock particles', *J. Geol.,* **40,** pp.443-451

Wadell, H. (1933), 'Sphericity and roundness of rock particles', *J. Geol.,* **41,** pp.310-331

Wadell, H. (1934), 'Shape determinations of large sedimental rock fragments'. *Pan-Am. Geol.,* **61,** pp.187-220

Wadell, H. (1935), 'Volume, shape and roundness of quartz particles', *J. Geol.,* **43,** pp.250-279

Walton, W. H. (1948), 'Feret's statistical diameter as a measure of particle size', *Nature, Lond.,* **162,** pp.329-330

Weibel, E. R. (1970), 'An automatic sampling stage microscope for stereology', *J. Microsc.,* **91,** pp.1-18

Weibel, E. R. (1979), 'Stereological Methods – Vol. 1'. Academic Press, London

Wentworth, C. K. (1919), 'A laboratory and field study of cobble abrasion', *J. Geol.,* **5,** pp.507-521

Wentworth, C. K. (1933), 'The shapes of rock particles: a discussion', *J. Geol.,* **41,** pp.306-309

Wentworth, C. K., Wilgus, W. L., & Koch, H. L. (1934), 'A rotary type of sample splitter', *J. Sedim. Petrol.*, **4,** pp.127-138

Whalley, W. B. (1972), 'The description and measurement of sedimentary particles and the concept of form', *J. Sedim. Petrol.*, **42,** pp.961-965

Whalley, W. B., & Orford, J. D. (1982), 'Analysis of scanning electron microscopy images of sedimentary particle form by fractal dimension and Fourier analysis methods', *Scanning Electron Microsc.*, **2,** pp.639-647

Whiteman, M., & Ridgway, K. (1988), 'A comparison between two methods of shape-sorting particles', *Powder Technol.*, **56,** pp.83-94

Whittaker, E. T., & Robinson, G. (1924), 'The Calculus of Observations'. Blackie, London

Williams, E. M. (1965), 'A method of indicating pebble shape with one parameter', *J. Sedim. Petrol.*, **35,** pp.993-996

Williams, J. C. (1968), 'The mixing of dry powders', *Powder Technol.*, **2,** pp.13-20

Winkelmolen, A. M. (1982), 'Critical remarks on grain parameters, with special emphasis on shape', *Sedimentology*, **29,** pp.255-265

Wylie, A. G., & Schweitzer, P. (1982), 'The effect of sample preparation and measuring techniques on the shape and shape characterization of mineral particles: the case of wollastonite', *Environ. Res.*, **27,** pp.52-73.

Young, B. D., Bryson, A. W., & Van Vliet, B. M. (1990), 'An evaluation of the technique of polygonal harmonics for the characterisation of particle shape', *Powder Technol.*, **63,** pp.157-168

Zahn, C. T., & Roskies, R. Z. (1972), 'Fourier descriptors for plane closed curves', *IEEE Trans. Comput.*, **C-21,** pp.269-281

Index